Lecture Notes in Computer Science 7493

Commenced Publication in 1973
Founding and Former Series Editors:
Gerhard Goos, Juris Hartmanis, and Jan van Leeuwen

Valeria Herskovic   H. Ulrich Hoppe
Marc Jansen   Jürgen Ziegler (Eds.)

Collaboration and Technology

18th International Conference, CRIWG 2012
Raesfeld, Germany, September 16-19, 2012
Proceedings

 Springer

Volume Editors

Valeria Herskovic
Poentificia Universidad Católica de Chile
Department of Computer Science
Vicuña Mackenna 4860, Macul, Santiago, Chile
E-mail: vherskov@ing.puc.cl

H. Ulrich Hoppe
University of Duisburg-Essen
Department of Computer Science and Applied Cognitive Science
Forsthausweg 2, 47057 Duisburg, Germany
E-mail: hoppe@collide.info

Marc Jansen
University of Applied Sciences Ruhr West
Computer Science Institute
Tannenstraße 43, 46240 Bottrop, Germany
E-mail: marc.jansen@hs-ruhrwest.de

Jürgen Ziegler
University of Duisburg-Essen
Department of Computer Science and Applied Cognitive Science
Forsthausweg 2, 47057 Duisburg, Germany
E-mail: juergen.ziegler@uni-due.de

ISSN 0302-9743   e-ISSN 1611-3349
ISBN 978-3-642-33283-8   e-ISBN 978-3-642-33284-5
DOI 10.1007/978-3-642-33284-5
Springer Heidelberg Dordrecht London New York

Library of Congress Control Number: 2012946543

CR Subject Classification (1998): H.4.1, K.3.1, H.5.3, K.4.3, H.3.4-5, J.1, H.2.8

LNCS Sublibrary: SL 3 – Information Systems and Application, incl. Internet/Web
and HCI

Typesetting: Camera-ready by author, data conversion by Scientific Publishing Services, Chennai, India

Printed on acid-free paper

Springer is part of Springer Science+Business Media (www.springer.com)

Preface

This volume constitutes the proceedings of the 18th Conference on Collaboration Technology (CRIWG 2012). Founded in 1995 in an Ibero-American context, CRIWG has become a significant international and inter-disciplinary forum for researchers and professionals in the area of collaboration technology and related fields such as CSCW, CSCL, and Social Media. CRIWG gathers contributions on groupware from a wide variety of academic perspectives, ranging from theory building and engineering approaches to innovative design and evaluation methods. The CRIWG conference series is supported by the Collaboration Researchers International Working Group (www.criwg.org), an open community of researchers.

CRIWG is held annually in culturally interesting places alternating between the Americas and Europe. It is organized in such a way as to foster lively exchange, critical discussions, and collaboration. The location of CRIWG 2012 is the Renaissance castle of Raesfeld in the Lower Rhine region of Germany with the University of Duisburg-Essen and Ruhr West University of Applied Sciences as local organizers. The previous five CRIWG conferences were held in Bariloche, Argentina (2007); Omaha, USA (2008); Peso da Régua, Portugal (2009); Maastricht, The Netherlands (2010); and Paraty, Brazil (2011).

This year's conference program comprised 9 full papers, representing consolidated research work, and 12 short papers reporting on work in progress. The review process was particularly intensive and careful, ensuring that each submission received at least four reviews. The spectrum of accepted papers gives an interesting account of the ever changing field of collaboration technologies: Eight papers deal with designing, facilitating, and analyzing technology-enhanced collaborative learning, three of these involving mobile support and two using gaming approaches. Three papers cover the new topic of social media analytics related to networked communities. Conceptual and design models of general groupware systems and technology mediated social interaction are the focus of four papers, whereas another four deal with formal modeling and technical approaches to describing and analyzing collaborative interactions, synchronization and concurrency control, and defining principles of group formation. Two papers are related to the specific application field of collaboration support in emergency scenarios. The international distribution of contributions is reflected in the following figures based on the affiliations of corresponding authors: Brazil (3), Canada (2), Chile (1), Germany (6), Greece (1), Israel (1), The Netherlands (2), Portugal (1), Spain (2), and Sweden (2).

The CRIWG 2012 program also featured two invited speakers who addressed different aspects of computational analysis of interactions in electronic communities. Whereas Ralf Klamma (RWTH Aachen, Germany) addressed general aspects of "community analytics" including experience from ongoing EU projects,

Dan Suthers (University of Hawaii, USA) focused on a theoretical framework and an interpretation scheme for collaborative learning data.

CRIWG 2012 was supported by a large number of collaborators. Particularly, we want to express our gratitude to all program committee members and reviewers, to the CRIWG steering committee, to our sponsors (see conference website), to our invited speakers, and to Anne-Marie Hussein and Adam Giemza for their organizational and technical support. Finally we thank all authors and participants for making CRIWG 2012 a valuable and memorable experience.

July 2012

Valeria Herskovic
Ulrich Hoppe
Marc Jansen
Jürgen Ziegler

Committees

Steering Committee

Pedro Antunes	Universidade de Lisboa, Portugal
Marcos Borges	Federal University of Rio de Janeiro, Brazil
Gert-Jan de Vreede	University of Nebraska at Omaha, USA
Jesus Favela	CICESE, Mexico
Jörg M. Haake	FernUniversität in Hagen, Germany
Stephan Lukosch	Delft University of Technology, The Netherlands
José A. Pino	Universidad de Chile, Chile
Carolina Salgado	Universidade Federal de Pernambuco, Brazil

Program Committee Chairs

Valeria Herskovic	Pontificia Universidad Católica de Chile, Chile
H. Ulrich Hoppe	University of Duisburg-Essen, Germany
Jürgen Ziegler	University of Duisburg-Essen, Germany

Program Committee

Pedro Antunes	Universidade de Lisboa, Portugal
Nikolaos Avouris	University of Patras, Greece
Nelson Baloian	Universidad de Chile, Chile
Lars Bollen	Universiteit Twente, The Netherlands
Angela Carell	adesso AG, Germany
Luis Carriço	Universidade de Lisboa, Portugal
Cesar A. Collazos	Universidad del Cauca, Colombia
Gert-Jan De Vreede	University of Nebraska at Omaha, USA
Dominique Decouchant	LSR-IMAG, Grenoble, France
Alicia Díaz	Universidad Nacional de La Plata, Argentina
Yannis Dimitriadis	University of Valladolid, Spain
Jesus Favela	CICESE, Mexico
Benjamim Fonseca	Universidade de Trás-os-Montes e Alto Douro, Portugal
Hugo Fuks	Pontifícia Universidade Católica do Rio de Janeiro, Brazil

Doctoral Consortium Chair

Stephan Lukosch   Delft University of Technology, The Netherlands

Organization Committee Chair

Marc Jansen   Ruhr West University of Applied Sciences,
                         Germany

Organization Committee

Adam Giemza   University of Duisburg-Essen, Germany
Anne-Marie Hussein   University of Duisburg-Essen, Germany

Table of Contents

Design Aspects in CSCL

Conceptual and Design Models for CSCW

Social Networks and Community Analytics

Formal Models and Technical Approaches

Mobile CSCL Scenarios

Emergency Scenarios

CSCL Scripts and Games

Computer-Supported Collaborative Drawing in Primary School Education – Technical Realization and Empirical Findings

Lars Bollen, Hannie Gijlers, and Wouter van Joolingen

University of Twente, Dept. of Instructional Technology, The Netherlands
{l.bollen,a.h.gijlers,w.r.vanjoolingen}@utwente.nl

Abstract. Self-constructed external representation, especially when embedded in peer inter-actions, are supposed to be beneficial in learning and teaching and can positively affect the course and type of reasoning for various reasons, e.g. by providing a ground for explanations and self-explanations, by helping to disambiguate learners' mental models of phenomena, by reducing working memory load, and by increasing and sharing the task focus. This paper reports on the results of research efforts in investigating conditions that are advantageous in collaborative drawing activities in learning scenarios for young students. We describe the design, technical implementation and empirical results of a study with 94 primary school students working on a collaborative drawing task in various conditions that include awareness information, prompting and scripted activities.

Keywords: external representations, collaboration, shared workspace, primary school education, scripted collaboration, awareness support.

1   Introduction

Drawings and sketches are a common method to present and illustrate phenomena, models and processes in learning and teaching. In primary school and already in pre-school education, drawings are used by teachers, on a chalkboard, by text-book authors or in multi-media learning materials. Even more, drawings can be used before children can read and write or later, thus constituting an early and universal way of illustration. But more than the mere reception of given graphical representations, the active creation of drawings and sketches is widely regarded as having the potential of being beneficial to learning [1]. In various studies, the students comprehension of scientific phenomena benefited from drawing activities in contrast to text-only conditions [2-4].

In addition to the beneficial use of drawing activities in educational contexts, it may bring forth its full potential when applied in collaborative scenarios [5]. In collaborative drawing settings, students are challenged to share their ideas and to disambiguate their conceptual understanding of the matter at hand.

V. Herskovic et al. (Eds.): CRIWG 2012, LNCS 7493, pp. 1–16, 2012.

In this paper, we will investigate on details in the conditions that may support and enhance collaborative drawing processes in learning contexts. Concretely, we explore the effects of contextual awareness information and the use of collaboration scripts on the process and the learning outcomes.

Awareness refers to students' perception and knowledge about a certain situation [6]. The general goal of awareness support in educational software is to increase students' awareness of specific situations or processes so that they can use this information to adjust their learning activities. Within the present study awareness support focuses on students' intermediate products, i.e. increasing students' awareness about specific objects and characteristics (arrows, labels) they did or did not include in their drawings.

By structuring and sequencing students' collaborative learning activities scripts shape the interaction and try to facilitate processes that lead to learning [7]. Basically, a script provides students with detailed and explicit guidelines about the task, and its successive subtask as well as the expected mode of collaboration within each subtask [8].

Domain-focused [9] as well as transactive dialogues [10] contribute to individual learning gains [11]. The transactivity of the dialogue refers to the extent students refer and build on others' contributions. An example of a dialogue excerpt with a relatively low level of transactivity is the externalization of a new idea. Dialogue excerpts that focus on the integration of a partners ideas in a students' own reasoning or critically discuss a partners contribution are considered highly transactive [7]. We assume that the awareness support improves what learners consider in their discourse (epistemological aspects of the discourse), and the script affects the level of transactivity.

As one major focus of this paper, we will deal with the technical requirements (and the subsequent implementation) that arose from the theoretically founded research questions and the following study design. We will point out, that the presented approach relies on the creation of "stable research prototypes" with the help of easy-to-use and easy-to-integrate frameworks and existing components.

In the following, we will elaborate on the details of the study setup, on the technical realization and on empirical findings. Section 2 explains the design of the study, its various conditions, planning in time and how data has been gathered with the help of questionnaires, observations, recordings and the produced artifacts. Section 3 will elaborate on the technical requirements that followed the design of the study and their respective implementations. As many software applications in the area of Technology Enhanced Learning can be regarded as highly specialized prototypes, which constitute a central aspect of the research in this field, we think it is vital to share our (technical) design decisions, implementation efforts, experiences and lessons learned in this respect. Section 5 will summarize our empirical findings, while section 6 concludes with a discussion and outlook.

2   Study Design

As a combination of the existing research results (as pointed out in the previous section) and our interest in the area of collaborative drawing activities in learning and teaching, we formulated the following research questions:

— To what extent do awareness prompts support students' discourse quality?
— To what extent does the collaboration script support students' discourse quality?
— To what extent do awareness prompts facilitate knowledge acquisition concerning the domain of photosynthesis?
— To what extent does the script facilitate knowledge acquisition concerning the domain of photosynthesis?

The basic idea for setting up an experiment that investigates into these questions, was to choose a challenging topic for primary school learners around age 10-11. Our choice fell to the topic of photosynthesis, since this is typically a new, unknown topic in this age group and presents rich opportunities for the creation of an explanatory drawing. Our experimental setup included three different conditions – one control condition with plan collaborative drawing activities, and two experimental conditions with awareness support or with scripted collaborative activities. In between the various phases of the experiment, different kinds of questionnaires (concept recognitions and open recall tests) were administered to the participants. Audio recordings of the learners' discourse and general observations have been gathered.

In details, the experimental setup consisted of

— an instruction and training phase, in which learners could work individually with the computer supported drawing environment on the topic of the "water cycle"
— a concept recognition test that presented learners with terms from that would relate to the topic of "photosynthesis" or not
— learning material that introduced the participants into the topic of "photosynthesis"
— an open recall test, which consisted of six questions that asked students to describe and explain specific aspects of the photosynthesis process.
— a control condition, that presented the learners with a collaborative, synchronized drawing tool
— an awareness condition, which included feedback prompts that helped the learners with hints about the quality of their drawing
— a script condition, that included a transition from an individual to a collaborative drawing phase

Table 1 shows an overview of the experiment planning. The row printed in bold letters denotes the phase where the participants actually were divided into the different conditions

Table 1. Experimental setup with timing and various conditions

Time / Condition	Control	Awareness	Script
35 min.	Instruction, training & practice		
5 min	First concept recognition test		
15 min.	Introduction to Photosynthesis		
10 min	Second concept recognition test and ffirst open recall test		
35 min.	Collaborative drawing phase	Collaborative drawing phase with additional aware-ness prompting	Individual phase, selection phase, discussion and collaboration
10 min.	Third concept recognition test and second open recall test		

3   Technical Realization

The above mentioned study design yielded a number of requirements to the technical setup to create a successful setup for the planned experiments to be successful.

3.1   Requirements

Hardware and Input Devices

To create activities that are based on known and established activities in primary school environments, we needed input devices that especially facilitate the computer-supported creation of drawings in a way that is as close to pen & paper as possible. Any deviation from this, e.g. using the computer mouse or a touch-based input method, is expected to cause unwanted bias in the learners' behavior and empirical findings.

Apart from the input device, computational devices were required as well to realize the foreseen scenarios. In some cases, the computational device and pen-based input device are realized together in one device, e.g. in tablet PCs or convertibles. In other designs, a pen-based input device is attached additionally to a computer or notebook, e.g. in the form of graphic tablets or interactive pen displays[1].

We could not rely on suitable equipment or a network infrastructure that would easily allow to implement the planned experiments in the targeted primary schools.

---

[1] Examples would be e.g. the tablet series from Trust (see http://www.trust.com for more details, last visited on April 12, 2012) or the Cintiq pen displays from Wacom (see http://www.wacom.com for more details, last visited on April 12, 2012).

Communication and Synchronization Infrastructure
The implementation of the foreseen collaborative drawing scenario required means for communication and synchronization between pairs of drawing applications. Apart from synchronizing the data model (the actual drawing), features like awareness support or transitions between phases in the scripted condition (as mentioned above) also called for mechanisms to execute remote commands in an RPC-like fashion. A convenient solution would allow both means of communication and data exchange.

The application scenario, i.e. setting up an experiment on collaborative, computer-supported drawing in several primary schools, suggested to build upon an ad-hoc, lightweight, and robust server solution. As a connection to a server outside the schools may not be easily possible, a solution that possibly resides together with the actual drawing application stood to reason. Also, to prevent a single point of failure, a communication infrastructure that works separately for each pair of computers seemed appropriate.

Drawing Tool
The drawing tool required to provide features for easy, stroke-based drawing activities. It needed to be as intuitive as possible to reduce the time needed for training and familiarization. Earlier experiences uncovered the need for a feature that would allow the learners to create boxes with textual input in their drawings. Certain objects or processes that need to be represented in a drawing are hard to visualize (e.g., in our case, "water vapor" or "sunlight"), so the young learners called for and appreciated the feature of adding short, textual descriptions to their drawings. Of course, the drawing tool was in need of features for data storage and logging to allow for extensive analysis subsequent to the actual experiment. Data storage and logging should be accomplished automatically and fully transparent to the user to avoid unnecessary distraction. The drawing application needed to be able to make use of various pen-based input devices, to flexibly react on different hardware configurations (tablets, pen displays, tabletPC etc.). If schools provided their own computer equipment, the utilized operating system may vary - typically between Microsoft Windows, Linux, or Mac OS. Thus, the drawing tool should be implemented in a platform-independent way. At last, the implementation of the drawing tool needed to be able to hook into the before mentioned communication and synchronization infrastructure.

Experimental Conditions
Overall, the combined choice of hardware, communication and synchronization infrastructure, and the drawing tool features needed to allow the realization of the experimental conditions mentioned in section 2.

The control condition consisted of the joint creation of a drawing between two learners in a shared workspace environment, following the "what you see is what I see" approach in a co-located classroom situation. Since the learners were able to directly speak to each other, no other means of computer-mediated communication was necessary.

The "awareness condition" added helpful hints in the form of pop-up prompts in comparison to the control condition to the drawing tool. The prompts were supposed to be partly content-sensitive with respect to the learner-created drawing. The awareness-mechanism was required to be able to detect separate objects in a drawing, to detect the use of text-boxes, and the occurrence of sketched arrows in the drawing (which are supposed to indicate processes in the applied domains of the water cycle and photosynthesis).

The "script condition" was planned to engage the learners in an individual drawing phase first, and would then switch to the creation of

Fig. 1. Wacom Cintiq pen display (taken from http://www.wacom.eu)

a joint drawing. Thus, the drawing environment needed to provide features for changing the drawing mode (from individual to synchronized), and to present an interface to pick elements from the individual drawings that would be used in the joint drawing.

The following sections describe the specification and implementation that followed the previously described collection of requirements.

3.2   Specification and Implementation

Pen-Based Display

As for the input device, we decided to choose Cintiq pen displays from Wacom (as shown in Fig. 1). These displays act as an external, second display and thus can be connect to most notebooks or desktop computers. They allow for a natural, pen-based usage of software applications. In comparison to touch-based interfaces (as known from various tabletPCs or tablets), one major advantage of this technology is that it only reacts on the pen, and especially not on a wrist that may rest on the display while drawing or sketching. Experience has shown that using these pen displays comes very close to the use of pen and paper. Also, we argue that using such kinds of pen display provides additional flexibility over other options, since they can be connected to any device that e.g. a school may provide or that will be used in future.

SQLSpaces Communication Infrastructure

As a result of the given requirements, the choice of the communication and synchronization infrastructure was decided towards a blackboard communication architecture [12]. As an example for a loosely coupled architecture, blackboard architectures have proven to be robust in case of irresponsive clients or slow, delayed connections, since clients are ignorant of each other's existence and would not be hampered by other clients malfunctions (if designed and used correctly). More concretely, we chose a TupleSpaces [13] approach as an instantiation of a blackboard architecture, and an implementation of this architectural approach called "SQLSpaces" [14-16][2]. The idea of TupleSpaces is to provide a conceptual framework to build a distributed system which is based on a client-server architecture and on the exchange of data that consists of tuples, i.e. ordered lists of primitive data (e.g. numbers, strings, boolean values etc.). SQLSpaces is one implementation of this approach, which allow clients to connect by using different programming languages, such as Java, C#, Prolog and more, which makes is highly suitable for implementing distributed systems that spread over different platforms and devices. In addition, SQLSpaces is able to be configured and to start up a server "ad hoc" at runtime, allowing for a flexible and quick solution for the use in experiments in schools. The rather quick and easy embedment of SQLSpaces in prototypical and experimental software developments (as it is often the case in the preparation of empirical studies in the area of Technology Enhanced Learning) makes it a suited candidate to realize a communication and synchronization architecture.

In SQLSpaces, clients can register callbacks that will be triggered on certain events, e.g. when tuples that match a given template are added, removed or modified. This feature eases the creation of shared workspace applications on the basis of replicated data models, i.e. each client and the SQLSpaces server hold a replica of relevant data (in our case: the learner's drawing).

Tuples are organized in so-called "spaces", which define a subset of all tuples stored on the server. In our case, it seemed natural to use one space to share the drawing data, and one space to exchange remote commands per pair or clients. Fig. 2 depicts the implemented communication and synchronization architecture.

Fig. 2 shows the separation of a shared, synchronized drawing and the execution of remote commands, which can be compared with an asynchronous remote procedure call, to enable two (or more) synchronized drawing application to behave similarly concerning e.g. awareness prompts or scripted behavior of the applications (see below). However, the exact protocol has to be implemented on top of the SQLSpaces framework. E.g., in the actual implementation, the following tuple would trigger the popup of a prompt with the given text in all clients connected to the same command space.

```
{ "command":String,    // indicating a command tuple
  "f81d4fae":String,    // the client ID
  "prompt":String,      // the type of command
  "Consider the use of arrows.":String} // com. property
```

SQLSpaces' characteristics of providing a light-weight server component, which can be started up in an ad-hoc manner at runtime, gave us the opportunity to use an SQLSpaces server for each pair of synchronized drawing tools. Doing so, we could distribute the load and overhead of synchronization over several notebooks, thus

---

[2] For detailed information, please visit http://sqlspaces.collide.info (last accessed on April 13, 2012).

avoiding bottlenecks in e.g. computational power or response times. Moreover, we prevented having a single point of failure that could possible crash a running experiment for a whole class.

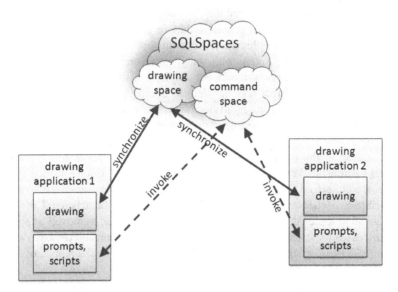

Fig. 2. Communication and synchronization architecture based on SQLSpaces

Java-Based Drawing Tool

The drawing tool, which has been build on top of the infrastructure described above, has been implemented in Java, since 1) we could rely upon and reuse components that had been created in earlier efforts in pen-based drawing scenarios [17-19], 2) an SQLSpaces client is available in Java, and 3) Java is platform-independent and thus would allow us to run the application on different operating systems that we might encounter in schools.

Fig. 2 shows the user interface of the drawing tool with a drawing created by two learners in the script condition (i.e. the collaborative creation of a drawing with awareness information for the learners). The toolbar to the left provides features for drawing and erasing strokes and for the creation of text boxes.

Additionally, the drawing tool contained features for automatically storing the created drawings as well as for writing detailed log files that would enclose the learners actions, including deleted elements of the drawing and text boxes, which would otherwise not be present in the finally stored drawings.

Scripting and Awareness Features

As for the advanced features of the drawing tool that were needed for the realization of the awareness and script condition, such as the recognition of drawing segments and shapes like arrows, and for the transition from an individual drawing phase to a joint drawing phase, the following implementation efforts have been made:

To recognize distinct objects in a drawing that consist of a collection of strokes, a naive Bayes distribution model is applied. This model is available in the data mining suite RapidMiner [20] and has been trained with data from previous studies with pen-based drawing applications. As a result, a drawing can be logically divided into segments, taking into account features like stroke creating time, stroke dimensions, location on the screen, pen pressure etc. In Fig. 3, the algorithm would recognize groups of strokes that belong to e.g. the sun, the tree in the center or figure to the house to the left. Subsequently, this information can be used to prepare context-sensitive feedback and awareness information to the learner about the progress of her drawing.

A second awareness information prompt based on characteristics of the drawing has been implemented by using the LADDER framework [21, 22]. This framework allows to specify shapes in terms of geometrical primitives, their characteristics and relations. In a stroke- or vector-based drawing, the LADDER framework is then capable of recognizing previously specified shapes. In our case, the use of arrows can be identified with a reasonable tradeoff between flexibility and accuracy.

In Fig. 3, you can see a popup-prompt that proposes the use of arrows, since the shape recognition did not find any arrows in the drawing. The awareness prompts were triggered by time, i.e. three minutes after the drawing activity started the tool would give hints on the (non-)existence of distinct objects in the drawing, after another three minutes the use of labels was checked, and after another three minutes the usage of arrows in the drawing was checked. If appropriate, prompts would appear synchronously on both students' screens.

The realization of the script condition, where the learners start to create an individual drawing, then decide which parts of both drawing would go into one joint drawing and then finalize it collaboratively, called for a suited interface to support the transition between the phases. This issue has been solved by introducing a split-screen interface, where the learners were able to pick the elements from their individual drawings to create one joint drawing in a drag-and-drop manner.

A screenshot from the split-screen interface for the transition from the individual drawings to the joint drawing is given in Fig. 4. The original, individual drawings can be seen to the upper left and lower left; by selecting elements of those drawings a new, merged drawing is created in the upper right. Selected elements of the individual drawings are moved (in a "cut-paste" fashion) from the left drawings to the joint drawing. To the lower right, instructions are given to the learner, including a button to indicate the end of this phase.

On the level of the communication and synchronization architecture, the individual drawings were organized in separate spaces on the SQLSpaces server, while the command space was still used to synchronize the behavior of both tools. Merged elements from the individual drawing were then copied to a third, synchronized drawing space (cf. Fig. 2).

Overall, our choices for using Wacom pen displays, an SQLSpaces communication architecture, a Java-based drawing tool and RapidMiner and LADDER to intelligently enrich the awareness and feedback functionalities, successfully lead to a stable, collaborative drawing environment that fulfilled the requirements that resulted from the original design of the experiment.

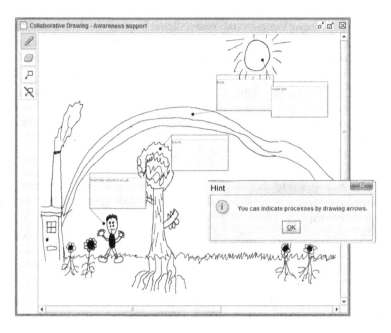

Fig. 3. Drawing tool with awareness prompt proposing the use of arrows

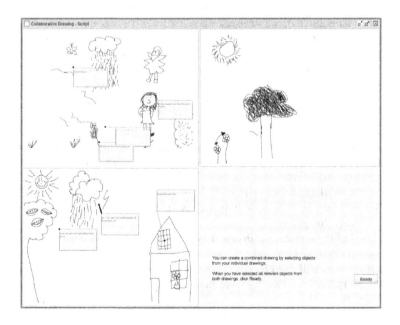

Fig. 4. Split-screen interface to create a joint drawing

4   Empirical Findings

The goal of the present study was to investigate to what extent awareness prompts and scripts support students' learning and interaction in a collaborative drawing setting. For this purpose a basic version of the collaborative drawing software was compared with two experimental versions in which students were supported with either awareness prompts or a script. In the present section we first report to what extent the supportive measures affected students' learning outcomes on both knowledge tests. Subsequently we report how the supportive measures affected students' discourse quality. And finally, we report on the relation between knowledge test and interaction measures.   Ninety-four fifth-grade students (47 dyads, aged 10-11), participated in this study, two dyads have been removed from the data set because they did not complete the entire learning session (resulting in 90 students, 45 dyads).

4.1   Knowledge Test

Two different tests (a concept recognition test and an open recall test) were administered to asses students' domain related knowledge. All means and standard deviations are presented in Table 2. Learning gains were calculated (intermediate test minus pretest scores, and posttest scores minus pretest scores) for all students. For the concept recognition test the results of an ANOVA revealed no significant effect of condition on the learning gain from pretest to intermediate concept recognition test $(F(2,87) = .110, ns)$. However, a significant effect of condition on the learning gain from intermediate test to posttest, $(F(2,87) = 5.533, p < .01$, Cohen's $d = 0.56)$ was found. In line with our expectations a post-hoc comparison of the means (using the Bonferroni procedure with adjusted alpha levels of .016 (.05/3) showed a significant difference in learning gains from intermediate to posttest between the control condition and the scripted condition in favor of the scripted condition.

Table 2. Mean scores and standard deviations on the knowledge tests

Condition	N	Concept recognition test			Open recall test	
		Pretest	Inter-mediate	Posttest	Pretest	Posttest
Control						
M	24	4.71	8.91	10.67	7.63	9.29
SD		3.24	1.71	1.27	2.12	1.92
Awareness						
M	34	4.56	9.15	11.65	7.53	10.67
SD		3.37	2.19	1.54	2.23	2.82
Scripted						
M	32	4.03	8.39	11.79	7.06	10.15
SD		3.16	2.46	1.22	1.65	1.64
Total						
M	90	4.41	8.81	11.43	7.38	10.01
SD		3.24	2.19	1.43	2.00	2.26

For the open recall test results of an ANOVA revealed a significant effect of condition (F(2, 87) = 5.449, p < .01, Cohen's d = 0.56). No significant differences were found between the awareness prompts condition and the scripted condition. A post-hoc comparison of the means (using the Bonferroni procedure) revealed that results were partially consistent with our expectations. Students in both experimental conditions outperformed their peers in the control condition on the open recall tests (all ps < .013). Contrary to our expectations, no significant differences on the knowledge test were found between the two experimental conditions.

4.2   Dialogue Quality

Students' discourse was recorded, transcribed and coded using a coding scheme based on the framework of Weinberger and Fischer [23]. Each utterance was coded on an epistemological and social dimension (both dimensions included codes for off task and paraverbal utterances). For students in the scripted condition the time they actually collaborated with their learning partner was shorter and they produced less utterances. Percentagewise scores were calculated for each student to rule out an effect of difference between conditions in amount of utterances. With respect to students' dialogue acts we were particularly interested in the percentages of epistemological utterances on concepts and processes and the level of transactivity of the dialogue. Since the awareness feature prompts students to draw concepts and processes we expected to find a higher percentage of concept and process related utterances for the awareness condition.. The results of a MANOVA with the percentages of utterances falling within the epistemological categories as a dependent variables and condition as independent variable revealed significant differences in the sub-categorical scores for the epistemic processes (F (26, 88) = 3.402. p<.001, Wilks' Lambda = .40, η^2 =.368). The results of subsequent ANOVAs followed by post-hoc analysis (using Bonferroni corrected alpha levels of .016) did not confirm our expectations. No significant differences between conditions for the percentage of defined domain related concepts and percentage of defined domain related processes were found. However, significant differences between conditions were found for the percentage of coordinative utterances (F (2, 87) = 6,386, p < .05, Cohen's d = 0.56) and the percentage neutral off-task interaction (F (2, 87) = 5,829, p < . Cohen's d = 0.56). Post-hoc Bonferroni corrected comparisons revealed a higher percentage of coordinative utterances in the script condition in comparison to the control condition and the awareness condition. The percentage of neutral off-task messages is significantly higher in the control condition than in both experimental conditions (all ps <.013).

Considering that the script encouraged students to compare, discuss and combine knowledge and ideas during the drawing activity, it was expected that students in the scripted condition would demonstrate a higher percentage of transactive processes (part of the social modes dimension) than the students in the awareness and the control condition. The results of a MANOVA with the percentages of utterances falling within the specific categories as a dependent variable and condition as independent variable showed significant differences (F (26, 88) = 3.402. p<.001, Wilks Lambda = 4.1, η^2 =.291) between conditions. Subsequent ANOVA's revealed differences

Table 3. Percentagewise scores and standard deviations on the epistemological and social dimension of the coding scheme (n=90)

	Control		Awareness		Script	
	M	SD	M	SD	M	SD
Epistemological:						
Concept naming	37.25	9.27	38.47	8.50	34.74	6.54
Concept definition	.47	1.01	1.78	2.16	1.53	2.01
Process definition	.040	.19	.32	.65	.45	.75
Concept process	2.42	1.76	2.73	2.66	3.77	2.41
Coordination	29.24	8.03	29.69	8.27	35.59	7.17
Off Task: Neutral	19.57	7.82	20.12	7.56	17.17	9.45
Off Task: Conflict	2.69	3.40	1.24	2.08	.66	1.04
Paraverbal	8.32	4.74	5.65	4.80	6.09	4.23
Social:						
Externalization	33.02	8.13	32.26	5.65	34.16	7.21
Elicitation	17.06	5.10	21.36	5.43	22.09	6.38
Quick agree	13.19	4.78	10.20	4.02	10.32	4.13
Quick disagree	3.42	3.28	1.47	1.83	2.41	2.95
Integration	1.77	2.44	4.27	3.51	5.31	3.87
Critical	.94	1.32	3.43	2.64	1.79	1.93
Off Task: Neutral	19.57	7.82	20.12	7.56	17.17	9.45
Off Task: Conflict	2.69	3.40	1.24	2.08	.66	1.04
Paraverbal	8.32	4.74	5.65	4.80	6.09	4.23

between conditions regarding the percentage of quick consensus building activities aimed at reaching agreement (F (2, 87) = 4,337, p < .05, Cohen's d = 0.67), the percentage of quick consensus building utterances revealing disagreement between students (F (2, 87) = 3,685, p < .05, Cohen's d = 0.48), the percentage of integration oriented messages (F (2, 87) = 7,623, p < .01, Cohen's d = 0.86) and the percentage of critical consensus building activities (F (2, 87) = 10,711, p < .01, Cohen's d = 0.43). Post-hoc (Bonferroni corrected) comparisons revealed that the percentage of quick consensus building activities (both agreement and disagreement oriented) was significantly lower in both experimental conditions. In contrast to our expectations results of the post-hoc comparisons show higher levels of transactivity (both integration oriented and conflict oriented consensus building) for students in both experimental conditions and not just for students in the scripted condition (all ps < .016).

In line with prior research [11] we expected that the transactivity of the dialogue was positively related to students' learning outcomes. This expected positive relation between transactive (integration oriented and critical consensus building) dialogue moves and learning outcomes was confirmed by the results of a stepwise regression analyses with the learning gains on the open recall test as the dependent variable. However, similar results were not found for the learning gains on the concept recognition test.

Table 4. Results of regression analyses with learning gain on the open recall test as the independent variable

	Model 1 β	Model 2 β	Model 3 β
Predictors			
Intergration oriented consensus building	.54*	.49*	.43*
Concept definition		.27*	.25*
Critical consensus building			.19*
R^2	.54	.60	.63
$R^{2\,Change}$ (df)		.06 (89)**	.03(89)*

*= p< .05   **=p<.01

Results were partially consistent with our expectations. Students in both experimental conditions outperformed their peers in the control condition on the concept recognition tests and the open recall tests and showed higher levels of discourse quality. Regression analyses revealed that highly transactive dialogue moves (integration oriented and critical consensus building) were significant predictors of students' scores on the open recall test.

5   Conclusion and Outlook

In the previous sections, we presented a study design, a technical implementation and empirical results in an experiment on collaborative drawing activities in primary school education. It could be shown that the chosen experimental setup along with the choice of input device, communication infrastructure, drawing application and supportive features such as awareness information and scripting support has been successfully applied in several school environments with 94 participants. Computer-supported drawing activities with pen-based input devices proved to be a promising and intuitive way to create self-constructed external representations, especially when embedded in peer interaction scenarios.

Empirical results have shown, partly in line with our expectations, the learning results and discourse quality can benefit from guidance in the form of collaboration scripts or awareness information in contrast to the creation of a joint drawing without these features. Students in the awareness condition or script condition had higher scores in concept recognition tests and open recall tests. condition engaged in more coordinative processes than their peers in the control and awareness condition. This is in contrast to finding of other studies that show that scripting alleviates the need for coordination [24]. Future research may reveal more details in the relation of other supportive features or combinations thereof in collaborative drawing activities.

Up to now, the detailed action log information have not been included in the current analysis. From these, we may derive additional information about if and how learning gains or discourse qualities relate to the students' *drawing processes,* e.g.

how many create/delete conflicts [16, 25] occurred or if there was a (im)balance in creating the drawing.

A simplified version of the presented collaborative drawing tool and other drawing and learning related tools are available from http://modeldrawing.eu.

Acknowledgements. We would like to thank Frank Leenaars, who build parts of the components that have been reused while implementing the drawing application and awareness support.

This work has been supported by the Stichting Kennisnet.

References

1. van Meter, P., Garner, J.: The Promise and Practice of Learner-Generated Drawing: Literature Review and Synthesis. Educational Psychology Review 17, 285–325 (2005)
2. Leopold, C., Leutner, D.: Science text comprehension: Drawing, main idea selection, and summarizing as learning strategies. Learning and Instruction 22, 16–26 (2012)
3. Schwamborn, A., Mayer, R.E., Thillmann, H., Leopold, C., Leutner, D.: Drawing as a Generative Activity and Drawing as a Prognostic Activity. Journal of Educational Psychology 102, 872–879 (2010)
4. van Meter, P.: Drawing construction as a strategy for learning from text. Journal of Educational Psychology 93, 129–140 (2001)
5. Brooks, M.: Drawing, Visualisation and Young Children's Exploration of "Big Ideas". International Journal of Science Education 31, 319–341 (2009)
6. Buder, J., Bodemer, D.: Supporting controversial CSCL discussions with augmented group awareness tools. International Journal of Computer-Supported Collaborative Learning 3, 123–139 (2008)
7. Weinberger, A., Fischer, F.: A framework to analyze argumentative knowledge construction in computer-supported collaborative learning. Computers & Education 46, 71–95 (2006)
8. Dillenbourg, P.: Over-scripting CSCL: The risks of blending collaborative learning with instructional design. In: Kirschner, P.A. (ed.) Three Worlds of CSCL. Can We Support CSCL, pp. 61–91. Open Universiteit Nederland, Heerlen (2002)
9. Anjewierden, A., Gijlers, H., Kolloffel, B., Saab, N., de Hoog, R.: Examining the relation between domain-related communication and collaborative inquiry learning. Computers & Education 57, 1741–1748 (2011)
10. Berkowitz, M.W., Gibbs, J.C.: Measuring the developmental features of moral discussion. Merrill-Palmer Quarterly 29, 399–410 (1983)
11. Teasley, S.D.: Talking About Reasoning: How Important Is the Peer in Peer Collaboration? In: Resnick, L.B., Saljo, R., Pontecorvo, C., Burge, B. (eds.) Discourse, Tools, and Reasoning: Essays on Situated Cognition, pp. 361–384. Springer (1997)
12. Buschmann, F., Meunier, R., Rohnert, H., Sommerlad, P., Stal, M.: Pattern-Oriented Software Architecture: A System of Patterns. John Wiley & Sons Ltd., West Sussex (1996)
13. Gelernter, D.: Generative communication in Linda. ACM Transactions on Programming Languages and Systems 7, 80–112 (1985)

14. Weinbrenner, S., Giemza, A., Hoppe, H.U.: Engineering Heterogeneous Distributed Learning Environments Using Tuple Spaces as an Architectural Platform. In: Proceedings of Seventh IEEE International Conference on Advanced Learning Technologies, ICALT 2007, pp. 434–436 (2007)
15. Weinbrenner, S.: Entwicklung und Erprobung einer Architektur für heterogene verteilte Systeme mit TupléSpaces und Prolog. Faculty of Engineering, Department of Computational and Cognitive Sciences. Diploma thesis. University of Duisburg-Essen (2006)
16. Bollen, L., Giemza, A., Hoppe, H.U.: Flexible Analysis of User Actions in Heterogeneous Distributed Learning Environments. In: Dillenbourg, P., Specht, M. (eds.) EC-TEL 2008. LNCS, vol. 5192, pp. 62–73. Springer, Heidelberg (2008)
17. Bollen, L., van Joolingen, W.R., Leenaars, F.: Towards Modeling with Inaccurate Drawings. In: Proceedings of International Workshop on Intelligent Support for Exploratory Environments at the International Conference on Artificial Intelligence in Education, AIED 2009 (2009)
18. Leenaars, F.: Facilitating Model Construction during Inquiry Learning with Self-Generated Drawings. Faculty of Behavioural Sciences, Department of Instructional Technology. Master thesis. University of Twente (2009)
19. van Joolingen, W.R., Bollen, L., Leenaars, F.A.J.: Using Drawings in Knowledge Modeling and Simulation for Science Teaching. In: Nkambou, R., Bourdeau, J., Mizoguchi, R. (eds.) Advances in Intelligent Tutoring Systems. SCI, vol. 308, pp. 249–264. Springer, Heidelberg (2010)
20. Mierswa, I., Wurst, M., Klinkenberg, R., Scholz, M., Euler, T.: YALE: Rapid Prototyping for Complex Data Mining Tasks. In: Proceedings of 12th ACM SIGKDD International Conference on Knowledge Discovery and Data Mining (KDD 2006), pp. 935–940. ACM (2006)
21. Hammond, T., Davis, R.: LADDER: A Language to Describe Drawing, Display, and Editing in Sketch Recognition. In: Proceedings of International Joint Conference on Artificial Intelligence (2003)
22. Hammond, T., Davis, R.: LADDER, a sketching language for user interface developers. Computers & Graphics 29, 518–532 (2005)
23. Weinberger, A., Fischer, F.: A framework to analyze argumentative knowledge construction in computer-supported collaborative learning. Computers & Education 46, 71–95 (2006)
24. Weinberger, A., Stegmann, K., Fischer, F.: Learning to argue online: Scripted groups surpass individuals (unscripted groups do not). Computers in Human Behavior 26, 506–515 (2010)
25. Bollen, L.: Activity Structuring and Activity Monitoring in Heterogeneous Learning Scenarios with Mobile Devices. Verlag Dr. Kovac, Hamburg (2010)

Training Conflict Management
in a Collaborative Virtual Environment

Katharina Emmerich, Katja Neuwald, Julia Othlinghaus,
Sabrina Ziebarth, and H. Ulrich Hoppe

University of Duisburg-Essen, Duisburg, Germany
{katharina.emmerich,katja.neuwald,
julia.othlinghaus}@stud.uni-due.de,
{ziebarth,hoppe}@collide.info

Abstract. In this paper we present a collaborative serious game for conflict management training in a role-playing scenario. The game *ColCoMa* (Collaborative Conflict Management) engages two players to participate in a conversation lead by an AIML chat bot mediator in a 2D virtual environment. Learning how to behave in conflict solving talks is supported by the separation of the game into a conversation phase and a reflection phase, causing players to change their perspective. Additionally, the learning process is emphasized by means of adaptive feedback based on individual analyses. Due to a multi-agent architecture approach, our implementation can be used as an easily adaptable framework for related collaborative learning scenarios.

Keywords: collaboration, multi-agent architecture, conflict management, serious games, role-play.

1   Introduction

Role-play is a vital instrument for a wide range of education and training scenarios, from children in schools to adults in organizational contexts [1]. Especially virtual role-plays have considerable advantages over real enactment [2]: Authentic simulated environments enable players to assume roles in particular contexts and to learn from the provided experiences given that these experiences are structured and well-designed [3]. Taking over roles allows people to explore new situations and to train how to act and react. Vice versa, observing role-play can lead to conclusions about one's own behavior [4]. Furthermore, virtual role-plays provide mobile, safe and repeatable environments for learners, as well as self-paced experiences that support the transfer of learning to the real world. Shaping the pedagogical outcomes poses a special challenge for virtual role-plays, as the effects usually depend on post-role-play reflection. Without feedback enabling the player to reflect on the role-play, the transfer to real-world situations cannot be ensured [1]

One pedagogically relevant application field of role-play is conflict management. Conflicts are at the core of all human interactions and thus also a part of everyday work life. Handling conflict the right way, especially in a working environment, has

V. Herskovic et al. (Eds.): CRIWG 2012, LNCS 7493, pp. 17–32, 2012.
© Springer-Verlag Berlin Heidelberg 2012

become an increasingly important social skill [5] and is also a subject of professional training. Advanced conflict resolution skills are especially crucial in facilitating peaceful, stable, and productive social interactions within communities [6].

This paper presents an attempt to train conflict management skills through a collaborative serious role-playing game based on psychological theories of conflict management. Through active re-enactment role-plays provide an added value when it comes to the training of successful conflict management. Since conflicts may have serious consequences if acted out in non-simulated environments, simulated virtual scenarios provide the possibility to explore, experience, and learn about conflict management and resolution without affecting the real world [6]. The aspect of collaborative learning is important in this context because it has the potential to improve social skills [7]. Our approach combines a collaborative environment for conflict enactment with the automatic generation of feedback. We believe that the affordances of role-playing games in combination with aspects of collaboration and intelligent computer-supported guidance and reflection will create a promising and innovative environment for training conflict management.

2   Related Work

Serious games are increasingly recognized as effective and powerful tools for facilitating learning and encouraging behavioral change [6]. The practical starting point for our considerations was the work of Malzahn et al. [8], who presented a framework for creating and conducting serious games, focusing on role-playing game-based learning scenarios in 3D environments. For demonstration purposes, they used a training scenario for apprenticeship job interviews. The distinctive feature of this approach is the connection of phases of immersion during the role-play and phases of reflection based on the assumption that reflection phases have to be adequately supported for successful learning. The following sections provide an overview of further serious games to be considered in relation to the three fields of interest: role-playing, conflict management, and collaboration.

2.1   Serious Games Related to Role-Paying

The use of role-playing is becoming prominent in serious games due to its positive effects on the learning of social skills [9]. There is a number of serious games based on role-playing covering a wide range of areas such as medicine, education, management, security, and engineering. In the following we focus on serious role-playing games for the training of social and operational skills related to the context of our work.

Choices and Voices[1] is a role-playing game in which players explore and discuss the issues and influences leading to tension and disruption in communities. The game provides two different scenarios, each making the player face a number of moral

---

[1] www.choicesandvoices.com

dilemmas, which ultimately determine the outcome for herself, her family, and her friends [10]. Although each player has the same choices to start with, everyone makes different decisions and thereby essentially changes the outcome of the game. *Choices and Voices* aims at creating an experience that acts as a catalyst and initiates discussions to prevent cruel and violent behavior.

Virtual Leader developed by *SimuLearn* [11] is a role-play based simulation program, which allows students to practice leadership styles and skills within a realistic 3D environment by participating in virtual business meetings with animated non-player characters[2]. It was designed to enable facilitators and educators to offer an engaging role-play to wider audiences, providing a safe and highly immersive opportunity to practice leadership skills like negotiation, collaboration, influencing, and conflict resolution.

Both systems represent structured approaches aiming at imparting operational competences based on role-playing and simulation using artificial intelligence. Interactive scenarios are integrated into a narrative in which players must make a range of decisions and consider different points of view. Each of the games provides meta-feedback on the player's decisions and behavior during the role-play, which is prompted at the end of each scenario to enable reflection. Both games succeed in representing experiences from which players may learn, although both are designed as single-user systems and therefore do not support collaborative learning. While both systems broach the issue of conflict resolution in some way, neither focusses on conflict management as such.

2.2   Serious Games Related to Conflict

While there are a number of approaches to the playful training of conflict scenarios, they typically focus on other aspects of the game environment. One prominent example, the interactive story *Façade* [12], for instance, is built around the idea of a completely unrestricted dialogue system, essentially granting the players the opportunity to communicate with non-player characters as they would with people in real life. The player has to try to resolve a conflict between a married couple by talking to the conflicting parties in order to find out where their issues stem from and give them some counseling in order to save their marriage.

A more recent approach is the conversational storytelling game *Office Brawl* [13], which also asks the player to act as mediator between two fighting parties. The idea behind this work was to create a game that supports emergent storytelling, essentially allowing players to experience a different game each time they play. To this end, *Office Brawl* makes use of bot-controlled NPCs with predefined action-reaction patterns.

While both *Façade* and *Office Brawl* feature a conflict scenario including two fighting parties and a mediator, neither approach is explicitly directed at training conflict resolution. The player does not receive any information on how to approach the

---

[2] www.simulearn.net

conflict and as she herself becomes the mediator in the situation there is no formal structure to the conversation nor any explicit feedback on how well the player has done other than the eventual resolution or escalation of the given conflict situation. Finally, neither game can be played collaboratively since there is no possibility to play as one of the two conflict parties. A true conflict management trainer would certainly benefit from such features in order to actually leave an impression on the player. In order to learn from a game, structure and feedback are important [1].

2.3   Serious Games Related to Collaboration

Many research approaches indicate the positive effect of collaborative learning, and the growing popularity of multi-user virtual environments has led to an increased interest in the use of collaborative technologies for learning [14]. Different types of collaborative virtual gaming environments have been applied in various contexts, supporting a multitude of pedagogical approaches. The main pedagogical benefits of collaborative game-based virtual environments lie in the provision of multiple perspectives, creating self-awareness of the learning process and thus making learning an interactive and authentic social experience [14].

One example for a collaborative serious game with role-playing elements is *Woodment*, in which different teams compete to lead their wood logging company, situated on an island, to victory [15]. Players have to collaborate with their team mates in order to perform better than the other teams. *Woodment* players may engage in different activities like exploring the island (e.g. looking for hidden learning content), managing the company, and leveling up (by answering questions collaboratively or on their own).

There are also collaborative approaches relating to conflict scenarios, e.g. the *Siren* project, a multi-user serious game concept that focuses on educating young people on socially and culturally suitable methods of conflict resolution [16]. It has been designed to support the teachers' role by generating conflict scenarios that fit the teaching needs of particular groups of children with varying cultural backgrounds, maturity and expertise levels as well as the learning outcomes as specified by the teacher. Players can be divided into groups that have to face a conflict situation together, each scenario containing different goals and obstacles to overcome.

Central to the two projects is the aspect of adaptability with regard to different players with different skills and knowledge. While both approaches support collaboration, also enabling direct communication between team members, neither one of them features intelligent support, such as individual feedback and reflection on the players' behavior. Thus, these systems are not designed to be played without some kind of teacher or trainer.

Our work aims at combining the advantages of the different approaches presented and discussed in this chapter in order to create a meaningful and structured system for training conflict management based on role-play, theories and best practices of conflict management, and collaboration.

3   Approach

Based on the idea to implement a collaborative conflict management training embedded into a role-playing scenario, we have designed the serious game *ColCoMa*[3]. Since there is no specific advantage of the 3D environment used by Malzahn et al. [8] over a 2D environment, we decided to use a 2D approach. In a graphically represented virtual chat setting, a pair of players conducts a conversation about a given fictitious conflict. The conversation is moderated by a chat bot acting as mediator, and follows the typical structure of mediation talks. The main goal is to come to a conflict resolution at the end of the conversation by showing appropriate and constructive behavior. Feedback is given at the end in order to support learning processes.

3.1   Gameplay

At the beginning of a new game session, the player has to log in. Among two players each is assigned a predefined role in a fictitious conflict scenario: Mr. Meier is working in a big software company as a member of the computer support hotline team and conscientiously takes much time for his customers. Mrs. Schmidt, his supervisor, registers Mr. Meier's very long call sessions that keep other customers on hold. She wants Mr. Meier to work more efficiently. After a negative appraisal of Mr. Meier's performance on the part of Mrs. Schmidt the situation escalates. The conflict and possible causes are conveyed by an introduction in terms of a picture story with short descriptive texts. Both players receive different illustrations, each from the perspective of the respective player's role, which should result in conflicting points of view. The scenario is deliberately kept simple and comprehensible, focusing on the main conflict and each person's feelings in order to support both immediate understanding and empathy with the assigned role.

The main game session after the introduction is subdivided into two parts: the mediation (conversation phase) itself and the reflection phase. During the mediation each player sees the representations of her two dialogue partners on screen, similarly to sitting opposite to the other conflict party and the mediator. By means of an integrated chat the players are able to communicate with each other and the mediator bot as well as evoke facial animations using common emoticons. Furthermore, it is possible to access a notepad and a help section in order to receive additional information on the scenario and the game controls. The conversation is either successfully finished if a conflict resolution is achieved or canceled if the bot notices that the conversation does not advance anymore or if one player leaves.

Irrespective of the way the mediation talk ends, it is followed by the reflection phase, in which participants leave their role and reflect on their behavior from a third-person perspective. First, players have the opportunity to directly exchange feedback among one another in a free feedback chat without the mediator and using their real names instead of their roles' names. Afterwards, they receive textual feedback regarding their overall performance in the conflict conversation, followed by a replay

---

[3] Collaborative Conflict Management.

session. Here, the whole conversation is shown again step by step, but it is augmented with individual feedback at certain points with respect to the player's behavior. A player will, for example, be criticized for interrupting her dialogue partner and praised for especially positive contributions.

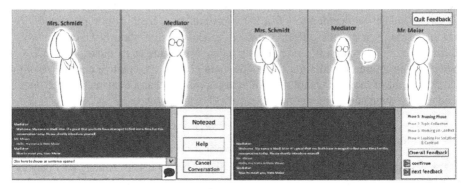

Fig. 1. Comparison of the graphical user interface in the first-person conversation phase (left) and the following third-person replay feedback session (right)

The separation of the reflection phase from the gaming phase is supposed to support the learning process as it advances meta-cognitive activities [8]. In the replay of the conversation role distance is reinforced by a change of the graphical interface: Now the player sees all three dialogue participants, including her own character. Thus a change from a first-person-perspective in the gaming phase to a third-person-perspective in the reflection phase is conducted (cf. Fig. 1), enabling the player to step out of her character. Players can navigate between single conversational phases, pause the replay or fast-forward to the next feedback marker. The conversational phases based on which the mediation is structured are pointed out in the following section.

3.2   Conversation Phases

Mediation talks typically consist of five conversational phases [17]:

1. Framing Phase
2. Topic Collection
3. Working on the Conflict
4. Looking for a solution
5. Contract

The Framing Phase at the beginning of the mediation talk is important to establish some rules for the participants' behavior towards each other throughout the conversation. In this phase, the actual conflict is not yet the focus. Rather than that, the involved parties are asked to state their personal hopes and goals for the mediation. Since the mediator has typically had a one-on-one talk with each party before, the actual mediation talk is mostly meant to help the conflict parties to reflect on their own positions as well as get an understanding of their opponent's point of view. To save time, in our game the rules are set by the mediator, who immediately goes on to ask the players for their goals regarding the mediation. This is a simple session of questions and answers, in which players statements are not further analyzed, except for general behavior, i.e. swearing or aggressive behavior are not tolerated.

In phase 2 the conflicted parties are asked to name topics they would like to discuss during their mediation. In this phase they are explicitly asked to not comment on the individual topics, but simply name them. They are asked for things that anger them or confuse them, regarding their relationship to the other party. In our scenario there are a few obvious topics, e.g. the performance review, working conditions, the parties' behavior towards each other, and their perspective within the company. In order for the mediator to recognize these topics, a list of keywords and phrases relating to each topic was generated. If a player mentions one of the topics or at least two keywords relating to the same topic, it is recognized by the mediator and added to the list of topics. To advance to phase 3, at least three topics have to be found. If players only manage to find two, the mediator will suggest a third topic. If they find less than two topics, the conversation will be terminated by the mediator for lack of contribution.

Phase 3 is at the core of the mediation talk. In this phase, the participants get to discuss each topic in detail with each party getting the chance to say why a topic is important to them, what they would like to change regarding the topic, and what they themselves could do to achieve this. In our case, the mediator chooses a topic, asks the first player about it, asks the second player about it, and then goes on to do the same with the next topic. Once the three topics have been talked about, the players are asked to put into words how they perceive their opponent's perspective now, after everything they have heard. The other player can then comment on whether that perception is correct, and, if not so, they may rectify their position.

Phase 4 is about finding solutions to the different topics that all parties can agree with, while in phase 5, if everything goes well, the conflict parties agree to adhere to the solutions they previously agreed upon and make a virtual contract. For the purpose of speeding up the game flow, phases 4 and 5 were merged together in our game with players being able to freely suggest solutions, which the other player then has to agree with in order for them to be added to the contract. Once players find enough solutions or do not have any more suggestions, they are then asked to both confirm that they are going to follow the contract from now on and can thus successfully finish the mediation.

3.3   Conversational Rules

Apart from the individual demands of each phase, conflict and mediation talks follow some general rules that players are asked to abide by. These rules include not swearing, not being aggressive or rude, and generally adopting an open and constructive attitude towards the conversation and the other parties. This includes sending more I- than You-messages, i.e. not being reproachful, not impairing the other person's autonomy. Furthermore, both conflict parties should try to name topics concretely, rather than vaguely hinting at what is bothering them, and should also be concrete when giving examples for behavior or situations they did not appreciate in the past. Although mediation talks are meant to deal with conflicts in a very objective way, showing emotions is not a bad thing as it makes a person more human and approachable. This can help the other conflict party to understand their opponent's perspective. Players are introduced to the conversational rules by the mediator at the beginning of the conversation. Keyword lists containing common swearwords and rude expressions are used to generate adequate feedback in case that one of the players uses them. Furthermore, during the reflection phase at the end of the game, players are provided with additional feedback on their behavior throughout the mediation talk, including explanations of which behavior is suitable.

4   Implementation

The technical implementation of *ColCoMa* entailed three main challenges: dialog modeling, programming of game logic and user interface of the client, and implementing a multi-agent architecture in order to keep components independent and to optimize performance due to parallel processing. We will discuss our approach regarding each of these aspects in detail in the following sections.

4.1   Dialogue Modeling

The conversational logic of the mediator bot was implemented using the XML-based Artificial Intelligence Markup Language (AIML). AIML is a common solution for AI-controlled passive chat bots, offering an easy syntax and a small number of control structures [18]. The dialogue flow was first specified in UML-like dia-grams and then transcribed into AIML files using *GaitoBot AIML Editor*[4]. The limited expressive capabilities of AIML required a number of "creative" work-arounds: For once, the passive nature of AIML-based bots did not suit our use case of a mediator guiding two conflict parties through a conversation. We therefore introduced keyword-based triggers to let the bot know when he had to become active, for example at the beginning of a new dialogue phase. The triggers and the bot's responses to them were predefined in the AIML files and fed to the bot from an external source. As is true for all artificial processing of natural language, it is very hard for AI to truly grasp the sense of what has been said, much less using the context of previous statements or the overall topic. For this reason, the general gist of each message was defined through a number of sentence openers provided to the users, who are thus able to freely finish the selected sentence. Each phase of the conversation provides the players with sentence openers indicating "affirmation", "rejection", and "further inquiry". Furthermore, specific sentence openers tailored to the demands of each phase are provided when needed, for example in phase 1 players have to explain their motivation for having the mediation talk and are thus provided with sentence openers like "I am here because...". Unfortunately, AIML is not made to deal with situations in which the bot

---

[4] www.gaitobot.de

has to differentiate between two players as is the case in our scenario. It is, however, crucial both for the flow of the conversation as well as for the individual user feedback that the bot knows which user is talking at a certain point of time. We therefore made the decision to provide each player with slightly different sentence openers, which are used to distinguish the two players or roles. We also expected the sentence openers to add to the overall atmosphere of the conversation, making it less stiff and more realistic. Depending on whether a player has previously been asked to speak or not, the bot is now able to scold players for speaking when it was not their turn or react accordingly to what the player has told him.

We further improved the basic AIML syntax by introducing feedback tags in the bot answers. The tags were used to mark situations in which the player is supposed to receive feedback during the reflection phase following the actual mediation. Tags include #Praise# for positive contributions to the conversation, #Interruption# for situations in which a player speaks without being asked to, #Repetition#, #Criticize#, and more. The tags are then processed and filtered out, so that players do not get to see the tags during the game.

4.2   Client Implementation

The complete client functionality was implemented in C# using *Microsoft XNA Game Studio 4.0*[5] as development framework, which facilitates the creation of games in combination with the programming environment *Visual C# Express*. The XNA framework is based on the native implementation of *.NET Framework 4.0* on *Windows* and provides a set of managed class libraries specific to game development, including functionalities for content and input processing, rendering, animation, sound and other components. The decision for using the XNA framework was based on the fact that it enables the quick development of a playable prototype. XNA provides the basic structure for an executable game, containing the main loop which allows the game to update the game logic and draw the game on the screen, as well as methods for initializing game components and loading content. A disadvantage of the XNA framework is the fact that it provides no standard GUI elements like textboxes or scrolling bars, but this disadvantage is compensated by the possibility to easily build up a simple interface based on sprites. With help of XNA it was possible to develop a working prototype, which includes a complete conversation simulation covering all conversation phases (cf. section 3.2).

4.3   Multi-agent Architecture

Besides the client, containing all user interface and game processing logic, there are ten additional program modules (agents) running independently based on a multi-agent blackboard architecture. The exchange of information between clients and agents is enabled by the use of *SQLSpaces* [19], an implementation of the TupleSpaces concept using a relational database. The client and all agents are writing and reading single

---

[5] Available at http://create.msdn.com/

tuples from the server without communicating among each other directly. This black-board architecture results in a loosely coupled and adaptive system: Agents may easily be adapted or replaced if demanded, for example if the scenario is changed. Furthermore, *SQLSpaces* facilitate the logging of all important game data of each playing session, which could be used for meta-analysis and comparison later.

All agents are implemented in C# and each is responsible for one task in the fields of dialog analysis, feedback creation or game control. The Register Agent is managing the log-in and arranges pairs of two players who have to play together. When a new client logs in, the Register Agent receives a request (callback) via the *SQLSpaces*. Every time two players have registered, the agent combines them as a pair, starts a new session and assigns the two player roles. In the process, information about the players' names and the assigned roles is logged in the TupleSpace, as the agent also considers the fact whether a player has participated in the game before and which role she has been assigned more often yet. According to that, the player can play repeatedly and trains each of the two roles.

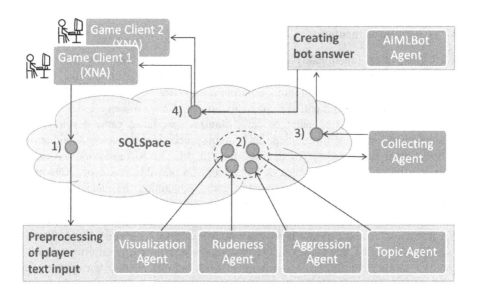

Fig. 2. Processing player input during the conversation phase

In the conversation phase of the game the players can send a sentence and receive an appropriate answer from the mediator bot. Several agents are involved in this process, as the player input is preprocessed before an answer is requested from the bot in order to generate the best fitting answer. On the one hand, this preprocessing helps to overcome the limited capabilities of AIML by analyzing certain aspects separately and thus reduces the structure of the AIML files. On the other hand, the creation of direct and adapted feedback is supported.

Fig. 2 shows the data-driven process in detail: When a conversation sentence is send by a player, the game client writes a tuple containing this sentence, the player name, the game session ID and a tuple ID into the TupleSpace (step 1). This tuple is read by the Visualization Agent, the Rudeness Agent, the Aggression Agent and the Topic Agent, which all in parallel analyze the player input regarding different aspects. The Visualization Agent searches for emoticons, that is to say short strings which express emotions, comparing a list of chat-typical emoticons to the player's input. If a match is found, the string representing one of nine facial expressions (e.g. smile, laugh...) is interpreted by the client and translated into an on-screen animation. The Rudeness Agent works in quite a similar way, comparing a list of swearwords and defamations to the player's input. The Rudeness Agent saves information on whether a player uses abusive language and how often she does so. Similarly, the Aggression Agent registers whether a player is being aggressive. This is operationalized in different ways: First, if the player writes more than one exclamation mark or solely uses capital letters (i.e. capslock), this is interpreted as screaming referring to common chat custom. Furthermore, a list of expressions related to the semantic field of threat and violence (e.g. 'kill', 'strangle'...) is compared to the player's input. The Topic Agent also analyses the user input but solely in phase two of the conversation. In this phase, players have to suggest topics and problems which they later want to discuss in order to solve the conflict. The Topic Agent searches for keywords and phrases in the players' input by means of a hash table: In sum there are four topics listed, each with a couple of related keywords. If one of these or the topic itself is mentioned by a player, the topic is recognized and stored for further processing.

Each preprocessing agent described above writes a new tuple containing the results of its analyses (step 2). These four tuples are then collected by the Collecting Agent and summarized in a single new tuple. This tuple signalizes that the preprocessing is done and hence activates a callback in the AIMLBot Agent (step 3). The AIMLBot Agent uses the *AIMLBot library*[6], which is a C# implementation of an AIML-based chat bot. The agent holds an array of bots – one bot for each active game session – and by means of the session number provided in the tuple the right one is chosen. Whether the bot receives the original player request or not is decided on the basis of the preprocessed analysis results: If a player is abusive or aggressive, her input string is replaced by an appropriate trigger ("swearword"/"aggression"), causing the bot to react rebukingly and then to repeat his last question again in order to obtain a proper answer. Otherwise, if a topic is mentioned in phase 2, the bot answer is created just on the basis of the topic instead of the original request, so that the bot can easily interpret and refer to the meaning of the player's input. In the case that none of those conditions is met, the bot receives the original string and answers according to the AIML pattern that matches the input. Finally, the AIMLBot Agent writes a new tuple, containing all the information that was collected so far and additionally the bot answer (step 4). This answer may also include the AIML feedback tags mentioned above in order to mark it as an important point for later feedback. The final tuple is read by both clients of the session, which in turn display the player input and the answer on screen.

---

[6] http://aimlbot.sourceforge.net

If no player is reacting for more than one-minute the Silence Agent writes a tuple which calls the AIMLBot Agent to react. The mediator bot then again asks the player to answer his question. This avoids an endless silence and leads to an abort of the conversation if the player does not react even after two more requests. The functionality of the Silence Agent is based on timers (one for each active session), which are reset every time a new conversation tuple is written into the TupleSpace, or otherwise evoke an event after one minute without player interaction.

The Me/You Agent is another agent which analyses the input of the players, but in this case not for the purpose of preprocessing, but just for the overall feedback that is created after the conversation. During the whole conversation, the Me/You Agent looks for and counts the amount of words referring to the speaker, like "I", "me" and "my", and words referring to the dialogue partners, like "you" and the names of the roles. This analysis helps to determine whether a player uses I-messages in order to express her own perception and feelings or rather makes suggestions about the other (cf. section 3.3). The result and an according estimation are integrated into the individual overall feedback after the mediation.

The creation of this overall feedback is the main task of the Feedback Agent. Besides the analysis results from the Me/You Agent it also grabs the counter information from the Rudeness Agent and Aggression Agent: The players receive commendation or criticism with respect to how many times they behaved inappropriately. The agent is activated by a tuple that is written into the TupleSpace at the time the mediator closes the conversation, composes the feedback for each player and makes the whole feedback available in a new tuple that is taken by the clients to show it on screen.

The last agent to be described is the Direct Feedback Agent, which is important for the last phase of the game, the replay of the conversation with integrated feedback. As mentioned above, we introduced tags to mark certain bot answers in the AIML code where later some kind of direct feedback for the player should be displayed. For example, if an answer does not fit the current question, the tag #Criticize# is used like this:

```
<category>
  <pattern>YES</pattern>
  <template>
    <condition name="activePlayer" value="all">
      #Criticize# What do you mean? My question cannot
      simply be answered with yes or no.
    </condition>
    [...]
  </template>
</category>
```

In this example, the created answer of the AIML bot to the player's input "Yes" is: "#Criticize# What do you mean? My question cannot simply be answered with yes or no." The bot answer then needs to be processed further; hence every tuple that is written by the AIML Bot Agent, that is to say every tuple containing a bot answer, is immediately taken by the Direct Feedback Agent. The agent searches for any feedback tag in the bot answer and removes it, so that players do not get to see the tags during the game. Thus the output on screen in the conversation phase would look like in Fig. 3a.

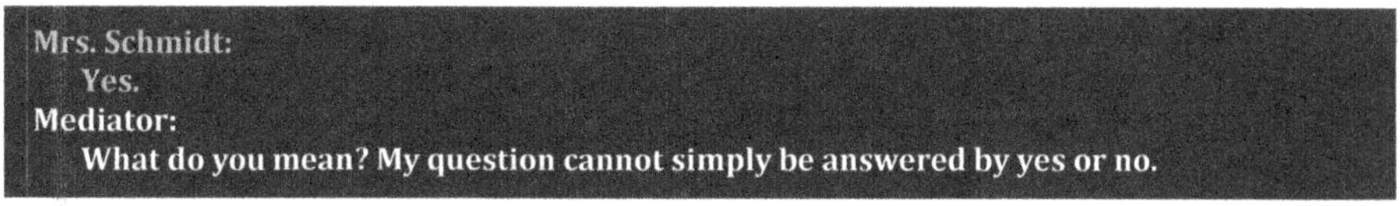

> **Mrs. Schmidt:**
>   Yes.
> **Mediator:**
>   What do you mean? My question cannot simply be answered by yes or no.

Fig. 3a. Output of the bot answer in the conversation phase: Feedback tags are removed

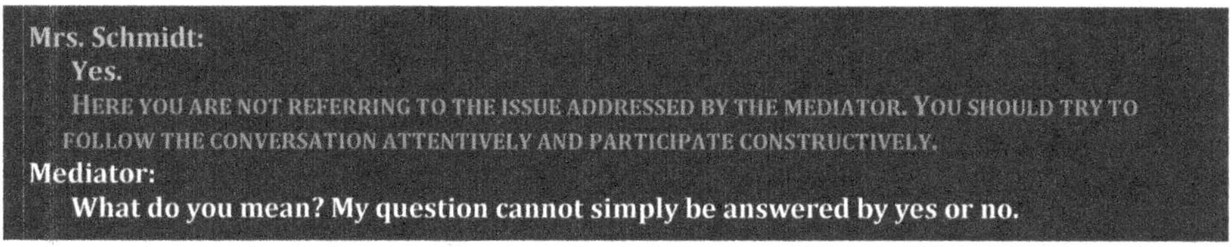

> **Mrs. Schmidt:**
>   Yes.
>   HERE YOU ARE NOT REFERRING TO THE ISSUE ADDRESSED BY THE MEDIATOR. YOU SHOULD TRY TO
>   FOLLOW THE CONVERSATION ATTENTIVELY AND PARTICIPATE CONSTRUCTIVELY.
> **Mediator:**
>   What do you mean? My question cannot simply be answered by yes or no.

Fig. 3b. Output in the reflection phase: Feedback is added according to the feedback tag

Internally, every tag is related to a couple of dedicated feedback sentences, from which one is randomly chosen. A new tuple is created, containing this feedback phrase as well as the information about the respective player and the ID of the conversation tuple. In the reflection phase, the client once again reads all conversation tuples and displays the sentences of the mediator and the players step by step in a replay. At the same time it checks out the TupleSpace for feedback tuples from the Direct Feedback Agent. Due to the ID of the conversation tuples, it can relate these tuples to the conversation tuples and thus show all feedback sentences at the right position, namely directly after the regarding player input that has to be commented. In the example case, a feedback related to the tag #Criticize# is "Here you are not referring to the issue addressed by the mediator. You should try to follow the conversation attentively and participate constructively.", thus the output in the reflection phase would look like in Fig. 3b.

5   Usability Test

A usability test in form of a user panel was carried out to get an initial feedback and ideas for further improvements. Eight students of the University of Duisburg-Essen participated in the test (50% male; average age: 25.78), playing the game in groups of two. Of the four groups, three got to play the game in the same location, while one pair played the game being located in different places. The players were observed during the game session and asked to "think aloud" while playing. Following the game including the reflection phase at the end, players were informally interviewed on an individual basis.

Generally, the game was received very positively with players noting that it was "fun" and that they felt they had learned something. Both the scenario including the introductory background story for each character as well as the characters themselves were unanimously perceived as appropriate, easy to understand, and universally applicable. The gameplay proved to be difficult for some players with some noting that they would have liked "more concrete instructions on how to use the interface". Problems included players not realizing that they had to select a sentence opener first in order to write something and players having trouble following the conversation due to the high rate at which bot answers were presented. A few players also noted that they were not always satisfied with the range of possible statements covered by the sentence openers offered to them. Furthermore, we were able to observe that players had the most trouble with the topic collection in phase 2, which required them to mention the right keywords in order for the bot to recognize a topic.

The feedback was immediately used to make a few changes to the client code as well as the agents. We implemented an additional screen introducing players to the interface right before the actual mediation phase. The list of topic triggers for phase 2 was extended by a vast number of keywords and phrases. The general speed of the conversation was reduced, so that new input was only shown after a short waiting period to give players time to finish reading the previous statements first. We hope to thereby facilitate players' interaction with the game and thus support their learning process.

6   Conclusion

In this paper we have shown an approach towards a collaborative, two-player 2D role-playing environment for the training of conflict management within a serious game. Our approach is centered around the idea of collaborative learning in an environment that adequately fits and supports the training of conflict talks. This is realized through the use of an immersive role-playing scenario, coupled with a separate reflection phase including individual player feedback. Feedback generation relies on a multi-agent architecture based on independent text analysis agents. What sets apart our work from others is the unique combination of a multi-player game supported by an intelligent chat bot and the adaptive feedback generation during the reflection phase at the end of the game. The idea behind this is that, through collaborative play, distributed players experience enhanced self-understanding and immersion. This is expected to effectively foster their conflict resolution skills.

One of the main challenges during the implementation was the realization of the mediator bot's behavior. This led us to the general question whether AIML is a sufficiently rich language for the realization of intelligent chat bots like the one in our scenario. While the XML-based structure of AIML is certainly easy to learn and use, the overall capabilities of the language are rather limited. For once, AIML bots are passive in their nature, working purely based on pattern matching algorithms. The only way to get the bot to actively contribute to the conversation is to use external triggers, still using pattern matching. Furthermore, AIML is not designed to support multi-user conversations, i.e. the bot cannot differentiate between two or more players or detect the sender of a message. Due to the use of sentence openers in our scenario, this problem could be worked around by introducing different sentence openers for each player. However, in real-life chats with free text input this would be a major issue. Furthermore, the order of pattern matching cannot be influenced. If the user input matches two or more patterns at a time, it cannot be predicted which answer the bot is going to use. Our multi-agent architecture enables a prioritization for which information the bot should receive about the player input. This problem could also be worked around through a pattern matching hierarchy within the actual AIML syntax; however, such a feature does not yet exist. One very basic flaw of AIML is the lack of operators or variables of types other than string, which are common in most programming languages. The availability of such structures would make it much easier to implement the bot's logic. All in all, there currently exists no solution for the easy implementation of intelligent chat bots comparable to AIML. However, the language itself could be improved by a broader palette of features and structures.

Our prototype can easily be tailored to other scenarios and thus serve as a basis for future research focusing on the adaption to other contexts and systems. Due to the flexibility of the blackboard architecture underlying the game environment, individual game components may easily be adopted, exchanged or adjusted to support similarly structured conversational games. Since the agents work independently of each other, the game architecture may serve as a framework for similar use cases and scenarios in the future. Other conflict scenarios could easily be implemented maintaining the general bot behavior and structure of the mediation talk. Additional feedback tags can easily be added, and topic and keyword lists expanded, altered, or changed vice versa. Moreover, the existing free feedback chat and replay session could be merged into one collaborative reflection environment. Thus players would be able to give each other constructive criticism directly relating to specific parts of the mediation talk, enriching the overall feedback by adding a real-life source.

References

1. Lim, M.Y., Aylett, R., Enz, S., Kriegel, M., Vannini, N., Hall, L., Jones, S.: Towards Intelligent Computer Assisted Educational Role-Play. In: Edutainment 2009, pp. 208–219 (2009)
2. Totty, M.: Better training through gaming. Wall Street Journal - Eastern Edition 245(80), R6 (2005)
3. Slator, B., Chaput, H., Frasson, C., Gauthier, G., Lesgold, A.: Learning by Learning Roles: A Virtual Role-Playing Environment for Tutoring. In: Lesgold, A.M., Frasson, C., Gauthier, G. (eds.) ITS 1996. LNCS, vol. 1086, pp. 668–676. Springer, Heidelberg (1996)
4. Martens, A., Diener, H., Malo, S.: Game-Based Learning with Computers – Learning, Simulations, and Games. In: Pan, Z., Cheok, D.A.D., Müller, W., El Rhalibi, A. (eds.) Transactions on Edutainment I. LNCS, vol. 5080, pp. 172–190. Springer, Heidelberg (2008)
5. Erdmüller, A., Jiranek, H.: Konfliktmanagement. Konflikten vorbeugen, sie erkennen und lösen. Haufe-Lexware GmbH & Co. KG, Freiburg (2010)
6. Cheong, Y.-G., Khaled, R., Grappiolo, C., Campos, J., Martinho, C., Ingram, G., Paiva, A., Yannakakis, G.N.: A Computational Approach Towards Conflict Resolution for Serious Games. In: Proceedings of the Sixth International Conference on the Foundations of Digital Games. ACM, New York (2011)

7. Turani, A., Calvo, R.: Sharing Synchronous Collaborative Learning Structures using IMS Learning Design. In: International Conference on Information Technology Based Higher Education and Training, pp. 119–129. IEEE Press, New York (2006)

8. Malzahn, N., Buhmes, H., Ziebarth, S., Hoppe, H.U.: Supporting Reflection in an Immersive 3D Learning Environment Based on Role-Play. In: EC-TEL 2010, pp. 542–547. ACM, New York (2010)

9. Greco, M.: The Use of Role-Playing in Learning. In: Connolly, T., Stansfield, M., Boyle, L. (eds.) Games-based Learning Advancements for Multi-Sensory Human Computer Interfaces, pp. 158–173. Information Science Reference, New York (2009)

10. Memarzia, M., Star, K.: Choices and Voices – A Serious Game for Preventing Violent Extremism. In: Yates, S., Akhgar, B. (eds.) Intelligence Management: Knowledge Driven Frameworks for Combating Terrorism and Organized Crime, pp. 130–142. Springer, Heidelberg (2011)

11. Knode, S., Knode, J.-D.: Using a simulation program to teach leadership. In: Proceedings of the 2011 ASCUE Summer Conference, pp. 86–92 (2011)

12. Mateas, M., Stern, A.: Façade: An Experiment in Building a Fully-Realized Interactive Drama. In: Game Developers Conference (GDC): Game Design Track (2003)

13. Glock, F., Junker, A., Kraus, M., Lehrian, C., Schäfer, A., Hoffmann, S., Spierling, U.: "Office Brawl" – A Conversational Storytelling Game and its Creation Process. In: Proceedings of the Eigth International Conference on Advances in Computer Entertainment Technology (ACE 2011). ACM, New York (2011)

14. Whitton, N., Hollins, P.: Collaborative virtual gaming worlds in higher education. ALT-J., Research in Learning Technology 16(3), 221–229 (2008)

15. Wendel, V., Babarinow, M., Hörl, T., Kolmogorov, S., Göbel, S., Steinmetz, R.: Woodment: Web-Based Collaborative Multiplayer Serious Game. In: Pan, Z., Cheok, A.D., Müller, W., Zhang, X., Wong, K. (eds.) Transactions on Edutainment IV. LNCS, vol. 6250, pp. 68–78. Springer, Heidelberg (2010)

16. Yannakakis, G.N., Togelius, J., Khaled, R., Jhala, A., Karpouzis, K., Paiva, A., Vasalou, A.: Siren: Towards Adaptive Serious Games for Teaching Conflict Resolution. In: Proceedings of ECGBL, pp. 412–417. IEEE Press (2010)

17. Proksch, S.: Konfliktmanagement im Unternehmen. In: Mediation als Instrument für Konflikt- und Kooperationsmanagement am Arbeitsplatz. Springer, Heidelberg (2010)

18. Wallace, R.: The Elements of AIML Style. ALICE AI Foundation (2004)

19. Weinbrenner, S., Giemza, A., Hoppe, H.U.: Engineering heterogenous distributed learning environments using TupleSpaces as an architectural platform. In: Proceedings of the 7th IEEE International Conference on Advanced Learning Technologies (ICALT 2007), pp. 434–436. IEEE Press, New York (2007)

Reusability of Data Flow Designs in Complex CSCL Scripts: Evaluation Results from a Case Study

Osmel Bordiés, Eloy Villasclaras, Yannis Dimitriadis, and Adolfo Ruiz-Calleja

GSIC/EMIC, University of Valladolid, Spain
{obordies,adolfo}@gsic.uva.es,evilfer@ulises.tel.uva.es,yannis@tel.uva.es
http://www.gsic.uva.es

Abstract. Several approaches have addressed the consistency and automatic enactment dimensions of CSCL scripts with data flow, but they have not appropriately tackled the problem of reusing such learning designs. For instance, workflow-based solutions such as LeadFlow4LD only capture particular case behaviors, instead of describing generic data flow situations. This limitation hinders the reusability of these designs because the workflow needs to be adapted for specific technical, teaching and social contexts. This adaptation is complex and time consuming, especially with a large number of students. In order to show the relevance of this problem, this paper analyzes the LeadFlow4LD approach through a real-world complex CSCL script. The study characterizes the reuse effort of CSCL scripts with and without data flow definition, in different social context settings. The findings illustrate how the data flow representation may affect the particularization of complex CSCL scripts, and pave the path for alternative, higher abstraction level representations of data flows, to reduce the reuse effort.

Keywords: reusability, data flow, complex CSCL scripts, workflow.

1 Introduction

Computer Supported Collaborative Learning (CSCL) scenarios promote the occurrence of effective interactions among learners by using scripting strategies [6]. Such effectiveness may be further enhanced by defining the data flow as a coordination mechanism that satisfies the dependencies established among learning activities [9].

However, existing Educational Modeling Languages (EML), such as IMS LD have a limited expressiveness to represent coordination mechanisms, and fail when giving support for a reusable and consistent definition of data flow for collaborative learning situations [13]. At different degrees, existing approaches addresses the consistency and automation perspectives through fixed data flow definitions. In fact, IMS LD-based data flows are implemented as "wired" links between data and learning activities [14]. On the other hand LeadFlow4LD (Learning and data Flow for Learning Design) proposes another approach. The

V. Herskovic et al. (Eds.): CRIWG 2012, LNCS 7493, pp. 33–40, 2012.

interoperability with IMS LD is granted separating learning flow and data flow, and the reusability dimension separating them into generic and instantiation level designs. [10]. Although the generic design is a "decontextualised" definition, the instantiated data flow depends on how real users or teams share tools, and how data and tools interact in specific technical, teaching and social context.

The reuse of such non-trivial CSCL scripts implies adequate group formation, tool and resource allocation for particular cases, but also the "how" and "what" to adapt in the data flow structural design [2]. For example, may consider as a learning goal within a peer review activity the perceiving of three artifacts instead one. In such case, two additional workflow activities should be incorporated to the original data flow definition, in order to meet the new requirement. This adaptation may be considered as a complex and risky task by teachers which are not familiarized with process modeling. Additionally, such scenarios characterized by several users and shared documents increases the error-proneness as well as the designer cognitive load [10].

The paper aims to confirm this hypothesis concerning the reuse of CSCL scripts in different contexts, when data flow is considered in the modeling phase. Specifically, its objective is to show the high effort related to the reuse in current approaches. This effort is illustrated through a case study and evaluated using a simple quantitative distance metric between the original script and the design that is produced after reuse. In these terms a real-world CSCL script is implemented following the LeadFlow4LD approach. The detection and analysis of the reasons that generate such a high reuse cost provides the requirements for a new alternative solution that is briefly explained in the paper.

The rest of the article is structured as follows: Section 2 further illustrates the reutilization problem through an illustrative case implemented following the LeadFlow4LD approach. Section 3 evaluates the reusability of the selected example. Then, section 4 discusses the findings and an alternative solution is briefly outlined. Finally, section 5 summarizes the conclusions and future research work.

2   A Case Study: Reusing a Complex CSCL Script

In the field of CSCL scripting, the reuse refers to "the possibility of adapting the functionality of pre-existing designs to meet certain technological and institutional frameworks, teaching profiles or social settings" [3]. The reuse reduces the time and the effort employed by teachers in the context of a community where novel solutions and good pedagogical practices are shared among their members. This section present the MOSAIC case study shows the complexity of modeling learning and data flows in real life CSCL scripts, when adaptation and coordination mechanisms are also embedded.

The original MOSAIC session [11, p. 286], formalized through the IMS LD language, was a blended scenario in which online and face-to-face activities were combined. The main instructional goal of the experience, performed by twelve PhD students and a Teacher (12+1), was the interactive creation of a conceptual map on the topics of *Grid Services* and *Service Oriented Computing* subjects.

In order to fulfill this objective, several group and individual activities formed the structural design of the learning flow. In the MOSAIC case, the participants initiated the experience engaged into a *jigsaw* strategy to solve a conceptual map from complementary contributions; and then arrived at a final consensus map in two additional steps that corresponded to a *pyramidal* structure.

In this study, the same script was enriched by applying a *group heterogeneity* adaptation mechanism with the objective of forming more likely heterogeneous groups. This script also included a peer review assessment process, which required automatic, successive and synchronous exchanges of documents among participants. In order to give technological support for the definition and automatic execution of the data flow, the case is implemented following the LeadFlow4LD approach. According to this solution the data flow is defined separate from the learning flow, using a workflow language in order to promote the consistency and the automation perspective. Finally, both processes are executed synchronously according a master-slave coordination model. The decision adopted about the case implementation not only has considered the advantage of the LeadFlow4LD from the competing approaches, but also their commonalities on reusing data flow designs. In this regard the evaluation results may be extrapolated to them in terms of the eventual adaptation effort.

Once the script has been designed including the data flow definition, the instantiation phase begins. This phase requires particularizing the generic design in order to enact the current collaborative learning situations as well as the supporting data flow. In that case, as procedural aspect, the teacher creates one learning flow instance *per* user, and one data flow instance *per* user or group that share tools following the iLeadFlow4LD specification [10]. The structural design of the data flow is set according to the original **contextual settings** of the MOSAIC script (labeled as [**CS**_0]) that can be briefly described as follows: 12+1 users, 9 learning flow activities, 51 workflow activities deployed in 4 different data flow situations and 556 lines of code of the instantiation design.

However, the reuse of such biased scenarios implies not only the change of several parameters in users, group forming and resources allocation terms, but also setting the corresponding data links among tools or activities and according to this, re-structuring the workflow-based design. In this regard, proportional workflow activities should be added, removed or re-configured to fit in specific data and learning activity relationship. This work is a very hard task for a teacher, since the effort required to re-contextualize the scenarios is very high and the educator needs to be familiarized with process modeling. It makes the reutilization of data flow designs a problem in real-world collaborative learning situations, such as the MOSAIC case.

The case has been implemented conforming to the architecture proposed in [10] and to IMS LD Level B[1]. The data flow design has been formalized using the notation BPMN[2] and GoogleDocs has been used as a learning tool through which the *conceptual maps* are created. The access to these documents is granted

---

[1] http://www.imsglobal.org/learningdesign/ldv1p0/imsld_infov1p0.html
[2] http://www.bpmn.org/

by using direct links. Similary, coordination resources responsible of synchronism between learning flow and data flow have been implemented as GET HTTP requests to specific workflow activities offered as REST resources.

3   Reusability Evaluation

Similarly as in the software engineering domain, the reusability of CSCL scripts may be affected by factors such as its usefulness, quality and the cost associated to the script adaptation [12]. In terms of quality, the script must be readable, testable and easy to be modified. Although, the missing "structural instantiation" in current data flow settings and the use of workflow languages increase the complexity of reuse, as well as the effort involved in the corresponding structural adaptation. Obviously, in this process the human factor also plays an important role, regarding the teacher profile and his understanding of the data flow model. In a previous work [2] this problem was analyzed from the modeling point of view; in this paper this hypothesis is tested in this section through a quantitative analysis.

The methodology adopted follows a descriptive research approach [5]. First of all, is estimated "how far" are the structural designs of data flow in alternative social settings (labeled as [CS_x]), with respect to a base scenario (label [CS_0]) in terms of complexity metrics. Such measurements mainly evaluate how comprehensible is a model for a person regarding objective factors such as *Size*, and subjective factors like his modeling competence [8]. In a second phase, the gathered data is analyzed in order to identify behavioral patterns and thus characterize reuse in complex CSCL designs. In theory, larger process models will be more difficult to understand, but this perception depends of the mental representation the user made of this process when adapting such models. Thus several levels of understanding may range from the IT designer point of view to the practitioner teacher perspective.

In our study, a teacher familiarized with CSCL scripting and workflow modeling domains adapted the structure of a data flow design described as "base scenario", in order to meet other alternative scenarios, with different contextual settings and data exchange logics (in the scenario [CS_5], see table 1). Shared data are defined and learning tools selected; hence the focus of attention is on providing the social setting information and to define the new supporting workflow structure of the data flow. The error-proneness of this process will depend on how complex is the structure to adapt, but also on the facility of the script to be understood by the teacher.

The complexity of the base scenario and the adaptation effort for each alternative scenario may be estimated as the "Size" [8] of the fragments modified in both scripts, the workflow structure and the data flow instantiation setting, during the particularization process. Although the traditional **size** metric rather easily illustrates the adaptation effort in objective terms (sum of activities added, removed or re-configured), our choice is conditioned by the implementation of the case where two different representations coexist: the graphical notation of the

Table 1. Changes of data exchange logics for the evaluation scenario [CS_5]. Implemented data flow situations represented with prefix S.

Data flow Sit.	Original exchange logic	Alternative logic
S11,S12,S13	Peer review among individuals	Rotated peer review among individuals (>2)
S2	Teams to individuals peer review	Idem.
S3	Teams to individuals peer review	Teams to all individuals peer review
S4	Inter-group peer review	Idem

process and the instantiation definition based on a markup language. Both are measured based on the *Number of Activity* (NOA) and *Lines of Code* (LOC) metrics respectively. However, we also recognize the limitations of this metric regarding in further studies more rigorous methodologies and the use of more abstract metrics.

4   Results and Discussion of Case Study

Table 2 summarizes the main findings related to the structural design modifications associated to each one of the scenarios proposed. Each row shows the size of the fragment operated during the adaptation or repurposing process and several aspects can be highlighted from the results. As it was analyzed in previous studies [2], the changes in social settings impact to some degree in the structural design of the data flow.

Small changes lead to noticeable changes in the structural design of the data flow, especially on group-based activities. Nevertheless there are cases, such as the scenarios [CS_3] or [CS_4], where the changes in the social context impact the structural design ranging from about 18% to 29% on the workflow sequence, and 33% to 58% on the instantiation design. But beyond these changes, the learning flow result also affected with modifications ranging from 31,6% to a 47%, where alternatives routes are added or removed. This reflects the case frequently encountered in the literature related with the "wired" implementation of certain adaptation mechanisms that follows the IMS LD approach [4]. Note the amount of changes required, even in the situations where the complexity of the overall MOSAIC design is reduced with six users less, as seen in the scenario [CS_4] (see table 2). Other changes made on data exchange logics (scenario [CS_5]) reveal the significant effect of the interaction between data and tools adapting the structural and instantiation designs by 19,6% and 34% respectively, mostly concerning the interactions with individual tasks.

All in all, it can be said that the complexity of the structural adaptation is not addressed, even if the complexity of isolated structural designs is measured accurately. The cognitive load associated to this process depends of the amount of changes, but also on what aspects should be subject of change (instantiation settings and structural design modification steps). For instances, the changes made in group sizes do not obey to the requirements formally specified in the

Table 2. Structural design modifications regarding the MOSAIC script deployment with five different social context settings. The **base contextual settings** are Users = 12+1, Number of learning activities = 9, Number of workflow activities = 51, Lines of Code of the instantiation design = 556.

Label	Contextual Changes	Structural design adaptations				
		Users	Modified learning activities	Modified workflow activities	Lines of Code (instantiation design)	Other findings
[CS_0]	Base	12+1	—	—	—	—
[CS_1]	An additional user	13+1	—	10 (19,6%)	37 (6,65%)	One copy of S1 should be repurposed (S5)
[CS_2]	A user less	11+1	—	8 (15,7%)	37 (6,65%)	—
[CS_3]	Three users less	9+1	6 (31,6%)	9 (17,6%)	184 (33,1%)	Addition of two alternative routes or more
[CS_4]	Six users less	6+1	9 (47,4%)	15 (29,4%)	326 (58,6%)	Conditional structure removed from S3
[CS_5]	Changing exchange logics	12+1	—	10 (19,6%)	189 (34%)	—

script, but to suggestions encountered in the bibliography. Teachers must be aware of these conditionalities, which rise the discussion about assistance to a properly CSCL script instantiation [1].

The findings exposed previously point out to the requirements that should be satisfied for the solution to the problem of reusing complex CSCL designs. The usefulness of reusing such designs is directly associated with the complexity of the data flow definition. This study is limited to one case study that only changes the social setting. Despite these limitations, the study reveals that contextualizing pre-existing designs is also complicated, due to the biased and *hardcoded* nature of the current data flow implementations. In such cases, the data flow situations might be defined using high-level representation in terms of *abstract data flow templates*.

These ideas have similarities with other approaches. Miao et al. understood that coordination mechanisms that support group-based CSCL scenarios, should be defined using representations with a higher level of abstraction than IMS LD or workflow languages [9]. It should also be possible to map these designs to corresponding executable models and would be intuitively understood and used by practitioner teachers. On the other hand, in the domain of scientific experimentation, the WINGS initiative [7] promotes the use of semantic workflow representation in order to address the reusability bottleneck associated to the limitations of the workflow models languages to support adaptation and evolution, especially with multiple parallel processes. Similar to computational workflows i.e. WINGS initiative, these templates must be defined as data-independent structures whose properties are described as constraints. These constraints may serve as a basis to generate both the data flow situations instances and executables data flow processes. Keeping these concerns separated the designer is not forced to set the data flow design to specific context and also alleviates the cognitive load of the teacher who deploys such designs; hence the reusability would be enhanced.

5   Conclusions and Future Work

The problem of reusing data flow designs for complex CSCL scripts has been described and its reusability level has been evaluated through the implementation of a relevant case study, which is deployed in several educational situations. Since specific reuse metrics are missing in the domain of complex CSCL scripts, we adopted the workflow process *size* metric as indicator of the reusability level, as it has been used in the field of software engineering.

The findings reveal how including data flow definitions in real-world CSCL scripts severely increases the reuse effort for the case of changes in social context. Recovered data demonstrates that even small changes in the social context may require quantitatively noticeable structural adaptations unwilling to be assumed by teachers. This problem is associated to several factors such as the gap between instantiation settings and corresponding structural adaptation of designs, the rigidity of the representation used to define data flow designs in the context of CSCL scenarios and the absence of specialized authoring tools. Further works will present other case studies that would cover a broader range of real life situations.

Future research work will focus on the design and development of workflow abstract templates as the proposed solution, which has been briefly introduced in the previous section. The workflow abstract templates may serve as intermediate representations able to be refined at different levels of the design generic, deployed and grounded. The development of suitable authoring tools will enable us first to show the expected advantages through the same enriched MOSAIC case study, and second to evaluate the capacity of teachers who are the target users of the tool to reuse and adapt CSCL scripts for their own needs.

Acknowledgments. This work was financially supported by a MAEC-AECID fellowship to the first author, as well as by the Spanish Ministry of Research and Innovation (TIN2008-03023/TSI), by the Spanish Ministry of Economy and Competitivity (TIN2011-28308-C03-02) and the Autonomous Government of Castilla and León, Spain, project VA293A11-2.

References

1. Alvino, S., Asensio-Pérez, Dimitriadis.Y., Hernández-Leo, D.: Supporting the Reuse of Effective CSCL Learning Designs through Social Structure Representations. Distance Education 30(2), 239–258 (2009)

2. Bordiés, O., Dimitriadis, Y., Alario-Hoyos, C., Ruiz-Calleja, A., Subert, A.: Reuse of Data Flow Designs in Complex and Adaptive CSCL Scripts: A Case Study. In: Daradoumis, T., Demetriadis, S.N., Xhafa, F. (eds.) Intelligent Adaptation and Personalization Techniques in Computer-Supported Collaborative. SCI, vol. 408, pp. 3–28. Springer, Heidelberg (2012)
3. Bote-Lorenzo, M.L., Hernández-Leo, D., Dimitriadis, Y.A., Asensio-Pérez, J.I., Vega-Gorgojo, G., Vaquero-González, L.M.: Toward Reusability and Tailorability in Collaborative Learing Systems using IMS-LD and Grid Services. Advanced Learning Technologies 1(3), 129–138 (2004)
4. Burgos, D., Koper, R.: Practical pedagogical uses of IMS Learning Design's Level B. In: SIGOSSEE 2005, pp. 1–8. Heerlen, The Netherland (2005)
5. Delorme, A.: Statistical Methods. In: Encyclopedia of Medical Device and Instrumentation, vol. 6. Wiley Interscience (2006)
6. Dillenbourg, P., Hong, F.: The Mechanics of CSCL Macro Scripts. International Journal of Computer-Supported Collaborative Learning 3(1), 5–23 (2008)
7. Gil, Y., Ratnakar, V., Kim, J., Antonio González-Calero, P.A., Groth, P., Moody, J., Deelman, E.: Wings: Intelligent Workflow based Design of Computational Experiments. IEEE Intelligent Systems 26(1), 62–72 (2011)
8. Mendling, J.: Metrics for Business Process Models. In: W. Aalst, J. Mylopoulos, M. Rosemann, M.J. Shaw, C. Szyperski (eds.) Metrics for Process Models: Empirical Foundations of Verification, Error Prediction, and Guidelines for Correctness. LNBIP, vol. 6, ch. 4, pp. 103–133. Springer, Heidelberg (2009)
9. Miao, Y., Burgos, D., Griffiths, D., Koper, R.: Representation of Coordination Mechanisms in IMS Learning Design to Support Group-based Learning. In: Lockyer, L., Bennet, S., Agostinho, S., Harper, B. (eds.) Handbook of Research on Learning Design and Learning Objects: Issues, Applications and Technologies, pp. 330–351. IDEA group (2008)
10. Palomino-Ramírez, L., Bote-Lorenzo, M.L., Asensio-Pérez, J.I., Dimitriadis, Y.A.: LeadFlow4LD: Learning and Data Flow Composition-Based Solution for Learning Design in CSCL. In: Briggs, R.O., Antunes, P., de Vreede, G.-J., Read, A.S. (eds.) CRIWG 2008. LNCS, vol. 5411, pp. 266–280. Springer, Heidelberg (2008)
11. Palomino-Ramírez, L., Bote-Lorenzo, M., Asensio-Pérez, J., de la Fuente-Valentín, L., Dimitriadis, Y.: The Data Flow Problem in Learning Design: A Case Study. In: Proceedings of the 2008 8th International Conference on Networked Learning, NLC 2008, Halkidiki, Greece, pp. 285–292 (2008)
12. Suri, P.K., Garg, N.: Software Reuse Metrics: Measuring Component Independence and its Applicability in Software Reuse. International Journal of Computer Science and Network Security 9(5), 237–248 (2009)
13. Vignollet, L., Ferraris, C., Martel, C., Burgos, D.: A Transversal Analysis of Different Learning Design Approaches. Journal of Interactive Media in Education (JIME), Special Issue: Comparing Educational Modelling Languages on the "Planet Game" Case Study (2), 1–10 (2008)
14. Villasclaras-Fernández, E.D., Hernández-Leo, D., Asensio-Pérez, J., Dimitriadis, Y., de la Fuente-Valentín, L.: Interrelating Assessment and Flexibility in IMS-LD CSCL Scripts. In: Proceedings of the 8th International Conference on Computer Supported Collaborative Learning, CSCL 2009, pp. 39–43. University of Aegean, Rhodes (2009)

Towards an Overarching Classification Model of CSCW and Groupware: A Socio-technical Perspective

Armando Cruz[1], António Correia[2], Hugo Paredes[3], Benjamim Fonseca[3],
Leonel Morgado[3], and Paulo Martins[3]

[1] Centro de Estudos em Educação, Tecnologias e Saúde, ESTGL,
Instituto Politécnico de Viseu, Campus Politécnico de Viseu, 3504-510 Viseu, Portugal
cruz.armando1@sapo.pt
[2] UTAD – University of Trás-os-Montes e Alto Douro,
Quinta de Prados, Apartado 1013, Vila Real, Portugal
ajcorreia1987@gmail.com
[3] INESC TEC/UTAD,
Quinta de Prados, Apartado 1013, Vila Real, Portugal
{hparedes,benjaf,leonelm,pmartins}@utad.pt

Abstract. The development of groupware systems can be supported by the perspectives provided by taxonomies categorizing collaboration systems and theoretical approaches from the multidisciplinary field of Computer-Supported Cooperative Work (CSCW). In the last decades, multiple taxonomic schemes were developed with different classification dimensions, but only a few addressed the socio-technical perspective that encompasses the interaction between groups of people and technology in work contexts. Moreover, there is an ambiguity in the use of the categories presented in the literature. Aiming to tackle this vagueness and support the development of future groupware systems aware of social phenomena, we present a comprehensive classification model to interrelate technological requirements with CSCW dimensions of communication, coordination, cooperation, time and space, regulation, awareness, group dynamics, and complementary categories obtained from a taxonomic literature review.

Keywords: CSCW, groupware, taxonomy, classification scheme, meta-review, socio-technical requirements, group process support.

1 Introduction

As systems and tools evolve and become more complex, it is much harder to evaluate them with high levels of completeness. Taxonomies provide a way to classify them according to their distinctive characteristics while establishing a basis for discussion and improvement. Commonly understood as "the science of classification", taxonomy is the assay of the procedures and principles of evaluation, whose terminological genesis is derivative from the words *taxis*, signifying arrangement, and *nomos*, meaning study [1]. Its focus relies on the intelligibility and schematic arrangement of the phenomena through taxonomic units arranged in a classification model or an hierarchical

V. Herskovic et al. (Eds.): CRIWG 2012, LNCS 7493, pp. 41–56, 2012.
© Springer-Verlag Berlin Heidelberg 2012

structure. For the specific case of systems and tools developed to support group work, several taxonomic approaches were presented in the literature, including technology-oriented or cooperative work dimensions. Partly, this diversity can be justified by the increased complexity with the emergence of new groupware systems, but it is also a reflection of a lack of adequacy and/or scope of existing taxonomies. Grudin & Poltrock [2] argued that CSCW research community leaned forward slightly on the fundamental frameworks developed by Mintzberg and McGrath in the 1980s, and more research is needed to fill the gap between social and techie domains [3], with a better understanding of the nature of collaborative work and the amount of technology features. This view is reflected by the CSCW acronym, which was coined to define two aspects considered then – as now – significant, cooperative work (CW) as social phenomenon that characterizes group work, and computer-supported (CS) in the perspective of collaboration technologies that support it [4]. Currently, CSCW involves nomadic work activities and comprises observable practices such as planning, intellectual co-construction, task management, playing, massively production, mechanical assembly, problem-solving and negotiation, which can be reflected in the 3C model [5]. Groupware, a "sibling" term of CSCW, refers to technology itself and is usually conceptualized as "computer-based systems that support groups of people engaged in a common task (or goal) and that provide an interface to a shared environment" to empower human interaction [5]. It provides a shared space for cooperation and enables awareness among group members, representing an outcome of CSCW research which encompasses sociological features of cooperative work in multiple forms and application fields (e.g., healthcare, learning, military training, tourism, among others). Therefore, these concepts are correlated and occasionally understood as synonymous.

Grudin & Poltrock [2] claim for an evaluation of technology in use on real scenarios (e.g., hospitals, museums and homes) towards a formal theory of CSCW to support new group dynamics with awareness and adaptive mechanisms to the context of labor. However, some difficulties arose to identify system requirements, taking into account the way people work in group, the influence of technology in their activities [6] – and consequently, problems in developing systems that would be based on those requirements. The lack of a standard set of collaboration dynamics and systems is one of the major gaps related to decomposition of collaboration processes in view of subsequent definition of system requirements and specification [7]. Task typologies unfilled in the literature have been applied by matching technology to tasks. Complementarily, collaboration is a phenomenon that can change over time and this fact implies a need to examine the articulation of cooperative work activities [4]. Thus, it would be useful to reformulate prior approaches used in the taxonomic models [8] and develop a classification model to accommodate new systems with increased complexity.

In this paper, we review multiple taxonomies that have been suggested to evaluate CSCW and groupware, ordering them chronologically according to the literature dimensions. Subsequently, a set of evaluation categories is proposed towards a classification scheme that aims to encompass the general requirements of collaboration technologies and cooperative work dynamics, addressing the problem of the lack of standardization. This was accomplished via a content analysis method by searching the main common categories of classification models suggested in the literature. The aim is to bring an holistic and evolutionary perspective highlighting taxonomic categories.

In addition to this introductory section, the structure of this article is subdivided by: a methodological approach used to collect taxonomies; a meta-review of taxonomies proposed to evaluate CSCW and groupware domains, followed by a segment dedicated to the analysis of results; a section with a schematic organization of socio-technical requirements that can support collaboration; and finally, a reflection section constituted by syntactical remarks and future research possibilities based on existing gaps.

2   Methodology

According to Weiseth *et al.* [7], there is a lack of a practical, holistic framework that may conduct organizations and other social entities in their effort to specify, evaluate and acquire collaboration systems that can support their needs. Verginadis *et al.* [9] claims that further research is required to develop or validate ontological structures of collaboration processes for recurring high-value tasks. Emerging collaboration tools require that CSCW research understand the current context, significant effects in society that unfold around successfully implementations [2]. Penichet *et al.* [10] argue that new evaluations are required to fit the current collaboration systems, and Schmidt [11] pointed to the need of a theoretical framework for analyzing or modeling cooperative work and specifying requirements of computer-based systems meant to support cooperative work. There is a need to consider the space within which CSCW research is conducted to create an overarching theory and taxonomy of CSCW and groupware [2]. Due to the lack of systematic reviews with amplitude to cover taxonomic approaches presented in literature, a meta-review process is presented based on the Kitchenham's guidelines [12] to summarize the current background.

The scope of this study relies in re-analyzing the literature references about groupware characteristics at a technological perspective, and collaborative work categories related to the human factors and task dimensions, with the aim to provide a systematic approach surveying and synthesizing prior research in CSCW [13]. Due to their systematic methodological nature, involving a discovery of theory through a data qualitative analysis, Inductive Analysis [14] and Theory for Analyzing [15] support a substantial portion of the present literature review process with a focus on the evaluation model that summarizes information and conveys taxonomic categories.

2.1   Research Questions

In the first methodical compass, a set of Research Questions (RQs) is defined to organize the main proposals of this study:

RQ1. Is a taxonomic background useful to CSCW scientific community? How can it be enhanced according to the technological and group work paradigm shift?

RQ2. How can we fill the gap of an unified taxonomy to CSCW and groupware?

RQ3. How should we distill the socio-technical requirements for cooperative work and groupware systems from literature integrating them into a comprehensive model?

Following the standard systematic literature review method proposed by Kitchenham *et al.* [12], RQ1 can be conceptualized into a temporal context, where last known taxonomic review was carried out by Bafoutsou & Mentzas [16], and other taxonomy-related approaches maintained a similar focus to justify classification categories [e.g., 7, 8, 10]. In this sense, we explore post-2002 references giving a different perspective compared to previous taxonomic review studies, and 'dig' the basics of previous taxonomies related with the main characteristics of group work and support systems. The second question (RQ2) addresses the problem identified in Grudin & Poltrock [2] study, where they argued that "much has been done since McGrath [17] on the nature of tasks performed by groups", and new technological implementations require that CSCW researchers understand the current work context to identify significant impacts on the societal domains. That is, we do not have an unified evaluation model to classify the holistic nature of CSCW research field, and this aspect can be a research opportunity for researchers to learn from the past taxonomic approaches and contribute to the understanding of emerging phenomena at a socio-technical perspective. The third issue (RQ3) relies on the identification of a suitable set of categories to classify socio-technical requirements from literature, launching a research agenda for future studies with a more empirical genesis in psychological, sociological, anthropological, among other domains that characterize the origins of CSCW field [2, 6].

2.2   Search Process and Selection Criteria

The phase of exploratory literature review follows a manual search process to retrieve a set of taxonomy-related scientific articles and books since 1984, not only for being an historical year for the origins of this field [6] but also for representing the course of two landmark studies perpetrated by McGrath [17] and Mintzberg [18], not diminishing similar studies. The set of articles, books and technical reports of our sample were selected by have been used as sources in other review papers [e.g., 2, 16, 19] and as to complement these previous studies with new references in order to achieve an holistic perspective on the taxonomic categories proposed in CSCW and groupware domains.

The selection of studies complied with the process is shown in the Figure 1. First, an inquiry was made for a set of refined search terms in Google Scholar, ACM Digital Library and Web of Science databases (see [20] for a detailed comparison), joining a word sequence constituted by the terms 'taxonomy', 'classification' and 'evaluation', and aggregating them with 'groupware' and 'CSCW'. The search by terms 'collaboration', 'cooperation' and 'coordination' were not totally applied in this study due to its common representation in distinct fields, which can expand this taxonomic universe in the future with a different search approach. In addition, a broad-spectrum reference list was collected and was read their title, abstract or in full, only in the cases in which taxonomic nature was not clear in the abstract. In the case of Google Scholar indexed references, citation count was a selection criterion to organize results as an important bibliometric indicator to this analysis, which was obtained from Google Scholar citation index. A review in the reference list of each publication allowed us to recognize taxonomic studies that were not found in the previous search.

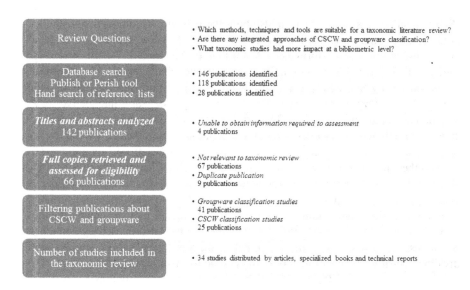

Review Questions	• Which methods, techniques and tools are suitable for a taxonomic literature review? • Are there any integrated approaches of CSCW and groupware classification? • What taxonomic studies had more impact at a bibliometric level?
Database search Publish or Perish tool Hand search of reference lists	• 146 publications identified • 118 publications identified • 28 publications identified
Titles and abstracts analyzed 142 publications	• *Unable to obtain information required to assessment* 4 publications
Full copies retrieved and *assessed for eligibility* 66 publications	• *Not relevant to taxonomic review* 67 publications • *Duplicate publication* 9 publications
Filtering publications about CSCW and groupware	• *Groupware classification studies* 41 publications • *CSCW classification studies* 25 publications
Number of studies included in the taxonomic review	• 34 studies distributed by articles, specialized books and technical reports

Fig. 1. Flow chart of taxonomic review process

2.3 Data Analysis

Focused literature review stage consisted in the full reading of papers. In this process, a scheme with the taxonomic attributes was developed to organize data about authors, year of publication, work dynamics in collaboration settings (communication, cooperation and coordination), temporal and spatial dimensions, group issues (types of group tasks, characteristics and size), technical categories of groupware applications (scalability, software and hardware), and complementary categories (e.g., usability). In this context, the afore-explained classification scheme supported a study of literature proposals based on the classification dimensions proposed by Bafoutsou & Mentzas [16]. To build this scheme, the previously chosen articles was read and extracted from them a set of core characteristics that were interrelated with reviews proposed in the past.

3   Diggin' the Literature to Find a Context: Taxonomic Anthology

As mentioned above, one of the earliest known taxonomic approaches for the study of groups was conceptualized by McGrath [17]. However, his research route was strongly influenced by significant preliminary studies. Carter *et al.* [21] classified tasks into six distinct typologies: clerical, discussion, intellectual construction, mechanical assembly, motor coordination and reasoning, which were introduced taking into account the group activity but cannot deal with the nature of the task outcome nor the relations between members at a coordination perspective. Meanwhile, Shaw [22] expressed the task complexity dimension, and McGrath & Altman [23] referred a need for systematic conceptual analysis of tasks and their relations to group members, within which the tasks could be classified according to: physical properties; behaviors needed and normally evoked by the task; behavior relationship among group members (e.g., cooperation requirements); and the task goal, criterion and outcome (e.g., minimizing errors).

Subsequent studies produced insights about the behavior requirements of intellectual tasks, specifying classification categories such as decision, production, problem-solving, discussion and performance. According to McGrath [17], the first really programmatic effort to show systematically the distinct characteristics of group tasks was carried out by Shaw [24], surveying them from past published studies of small groups and extracting six categories along which group tasks vary: intellective versus manipulative requirements; task difficulty; intrinsic interest; population familiarity; solution multiplicity or specificity; and cooperation requirements. Driven by the afore-referred visions, McGrath [17] extract the main ideas from literature and fit them together into a conceptually interrelated set of classification dimensions about tasks. The result was a group task circumplex constituted by four quadrants (generate, choose, negotiate, or execute), within which are specific task types: planning, creativity, intellective, decision-making, cognitive conflict, mixed-motive, contests/battles, and performance. In an organizational viewpoint, the model proposed by Mintzberg [18] claims that executives (*strategic apex*), managers (*middle line*), support staff, individual contributors (operating core) and people formulating work processes (*technostructure*) often have different approaches, constraints, opportunities for action or competing priorities [2].

In the groupware domain, one of the first approaches was proposed by Bui & Jarke [25], with a strong focus on communication requirements for Group Decision Support Systems (GDSS). In this approach, a group communication arrangement can be classified with respect to the spatial distance between decision meetings (remote and co-located), temporal distance of the participants (simultaneous and different time), control centralization (democratic or hierarchic), and cooperation degree in meeting settings (cooperation and negotiation). The contingency perspective for GDSS research [26] has three dimensions: i) task type (planning, creativity, intellective, preference, cognitive conflict or mixed motive), ii) member proximity (face-to-face or dispersed), and iii) group size (smaller or larger). Task type distinction was based on McGrath's group task circumplex [17], a vision corroborated by Jelassi & Beauclair [27].

Nunamaker *et al.* [28] proposed a taxonomy that replaced the task type criterion by time. Their scheme also takes into account the number of members of a team or group according to physical proximity and time dispersion. The time/space matrix was suggested by Johansen [29] with focus on the temporal and spatial dimensions. It is mainly concerned with the technological support for group activities, and has four classes relating the time and place of team members' interaction. Grudin [6] brought an extension that adds predictability of place and time to the original matrix.

The taxonomy proposed by Ellis *et al.* [5] highlights the significance of 3C model (communication, coordination and collaboration) in the group interaction support. In their opinion, collaboration is based on information sharing, and coordination is concerned to shared objects access. Similar taxonomies (e.g., [30, 31, 32]) were proposed to interrelate a set of application domains into the 3C model [30], where an evaluation based on modes of collaboration is organized by communication, information sharing and coordination categories, and the collaboration synchronicity degree (real-time or asynchronous) [32]. Ellis & Wainer [31] developed a functional decomposition to the groupware systems taking into account the three dimensions intended to developers whose focus lies on the user interface, specifically: the groupware ontology associated to the data structure; a coordination model that describes the effective management of interaction flow; and an user interface model to help the interaction between the users and system. In the same year, Malone & Crowston [33] proposed a taxonomy of coordination tools, where the focus relies on the supported management process and it is dedicated to management issues, fairly independent of technical features.

Hybrid taxonomies were introduced to associate categories of the central schemes (time/space, 3C model, and application domains), and add new taxonomic elements to categorize the technological and social components of CSCW research. In the taxonomy proposed by Jarczyk *et al.* [34], collaboration systems were characterized by the follow classes of criteria: functional, application, technical, usability, ergonomics and scalability. Mentzas [35] classified the coordination-related aspects of group technology into five categories: coordination model characteristics, type of processing, decision support, organizational environment, and objectives. Completing the evaluative review of Bafoutsou & Mentzas [16], new taxonomies was introduced to provide: i) a categorization of collaboration tools according to the underlying technology [36], ii) task (decomposition and complexity), group (size, composition, leadership, member characteristics, and subject) and technology (task support, tools, mode of communication, process structure, and design) [37], or iii) application-level categories of collaboration tools (e.g., workflow and group decision support) or collaborative services [16]. Basically, the hybrid taxonomies proposed in the literature can give a broad-spectrum classification perspective, integrating the main previously contributions to help programmers, academics and general public to understand collaboration systems.

4   Meta-analysis Results

As mentioned above, Bafoutsou & Mentzas [16] reviewed prior taxonomies presented in literature. Nevertheless, this review left out some studies (e.g., [11, 17, 18, 30, 25, 27, 28, 31, 32, 37, 38]) taking into account the time interval of analysis (1987-2002). The distribution of dimensions is represented within a classification scheme (Table 1), which is supplemented with more categories and taxonomies identified using our own methodical process. The sample of this review is circumspect to the 1984-2009 interval, where the selected criteria is based on the importance of McGrath's research into group tasks domain, and the complementarity to previous studies with a current systematic analysis. The review scheme is constituted by: time/space (which is the most addressed in the Table 1, where collaboration can be synchronous and asynchronous, as well as co-located and remote); CSCW characteristics (based on 3C model); group issues (size, characteristics and task types); technical criteria (scalability, software and hardware); and complementary features (e.g., ergonomics and usability, awareness, or application domains). The CSCW characteristics (such as cooperation, coordination, communication, articulation work, division of labor, among others) are at a similar level of group issues. In fact, they are included in taxonomic proposals since 1991. By a chronological distribution of literature, it can be seen that the interest in 3C model characteristics has grown in the last years. The attention on other aspects was stable, with clear exceptions to group issues, whose interest experiences a notable fall in the last years. Also, a concern with human-computer interfaces is really obvious, with this feature being an important issue. Starting on their specific work, we will take it a step further, adding the dimension of CSCW characteristics, bibliometric indicators, and including both the more recent literature and that they left out from their study. Our contribution is highlighted in bold, giving a perspective of unexplored taxonomies. The study presented here could be expanded by considering aspects such as 'collaboration patterns' [39], collaboration needs adapted from Maslow's hierarchy [40], and classification dimensions suggested by Boughzala et al. [41] to evaluate collaborative work and technology support centered on the MAIN+ method, which were not taken into account in the present version but offer improvement possibilities.

Table 1. Distribution of classification dimensions across the literature

Year	Author(s)	CSCW characteristics	Time/ Space	Group issues	Technical criteria	Other	Bibliometric indicators[¥]
1984	McGrath [17]			*			2798 citations
1984	Mintzberg [18]			*		Organizational Structure	8393 citations[§]
1986	Bui & Jarke [25]		*			Mode of Interaction	52 citations
1987	DeSanctis & Gallupe [26]		√	√			1747 citations
	Jelassi & Beauclair [27]		*			Mode of Interaction	71 citations
	Stefik et al. [42]				*	Development/HCI	1033 citations
1988	Kraemer & King [43]			√	√		494 citations
	Johansen [29]		√				801 citations
1991	Ellis et al. [5]	*	√			Mode of Interaction; Application-level	2912 citations
	Nunamaker et al. [28]		*	*			1590 citations
1992	Jarczyk et al. [34]			√	√	Mode of Interaction; Usability/Ergonomics; Application-level	8 citations
1993	Mentzas [35]	*	√		√	Mode of Interaction	32 citations
1994	McGrath & Hollingshead [44]			√			646 citations
	Grudin [6]		√				934 citations
	Malone & Crowston [33]	*		√			2589 citations
	Ellis & Wainer [31]					Development/HCI	228 citations
1995	Coleman [45]					Application-level	132 citations
1997	Grudin & Poltrock [32]	*	*				76 citations
1998	Fjermestad & Hiltz [37]			*	*	Mode of Interaction; Usability/Ergonomics; Organizational Structure	596 citations
2000	Ellis [36]			√		Application-level	22 citations
2000	Ferraris & Martel [38]	*				Regulation	27 citations
2002	Bafoutsou & Mentzas [16]		*			Application-level	134 citations
2002	Pumareja & Sikkel [46]	*	*	*		Application-level; Awareness Indicators	5 citations
2002	Andriessen [47]	*		*		Organizational Structure	6 citations
2003	Bolstad & Endsley [48]	*	*		*	Mode of Interaction	17 citations
2004	Neale et al. [49]	*					165 citations
2006	Weiseth et al. [7]	*			*		23 citations
2007	Okada [50]	*	*	*		Awareness Indicators	0 citations
	Penichet et al. [10]	*	*			Application-level	28 citations
	Elmarzouqi et al. [51]	*	*			Development/HCI	6 citations
2008	Mittleman et al. [8]		*		*	Application-level; Awareness Indicators	17 citations
2009	Giraldo et al. [52]			*		Development/HCI	0 citations
2009	Golovchinsky [53]	*	*				26 citations
2009	Briggs et al. [54]			*	*	Application-level	11 citations

√ Bafoutsou & Mentzas [16] classification dimensions

* Our systematic review's contribution

¥ Obtained from Google Scholar citation index at 8 April 2012

§ Number of citations related to the first known Mintzberg's approach

Bolstad & Endsley [48] proposed a classification scheme for collaboration tools intended to support the development of technology and acquisition of material for military purposes (i.e., face-to-face, video/audio conferencing, telephone, network radios, chat/instant messaging, whiteboard, program/application sharing, file transfer, e-mail, and domain specific tools. Moreover, categories are classified as collaboration characteristics (collaboration time, predictability and place, and degree of interaction), technology characteristics (recordable/traceable, identifiable, and structured), information types (emotional, verbal, textual, video, photographic information, and graphical/spacial), and collaboration processes (data distribution, gathering, scheduling, planning, tracking, document creation, brainstorming, and shared situational awareness).

Then, Neale *et al.* [49] proposed a pyramidal scheme for the evaluation of the support provided by collaboration systems to activity awareness. This taxonomy focuses some of the core CSCW characteristics (communication, coordination, collaboration, cooperation, information sharing, and light-weight interaction), linked to contextual factors, distributed process loss, work coupling, common ground, and activity awareness). Subsequently, Weiseth *et al.* [7] suggested a wheel of collaboration tools as a framework constituted by the collaboration environment, process and support, related to the functional areas of coordination (mutual adjustment, planning, standardization), production (mailing, search and retrieval, capturing, authoring, publishing), and decision-making (survey, query, evaluation and analysis, reporting, choice).

In 2007, Okada [50] introduced an hierarchical taxonomy to classify collaboration. From basis to top, it has: coexistence (place and time), awareness (influenced by human, spatial, and temporal factors), sharing (views, opinions, knowledge, operations, and others), and collaboration (cooperation, and assertion). According to the author of this study, the degree of assertion and cooperation shown by the group members influences the outcome of collaboration. Only high levels of assertion and cooperation result in coordination, a higher level of assertion results in collision, and if cooperation is higher than assertion the outcome is concession. Penichet *et al.* [10] considered that most of the taxonomies existing at the time were inadequate to classify more complex systems that include a large variety of tools. They argued that some tools are forced to fit in one category. In fact, these systems can be used in different ways and contexts in a synchronous and asynchronous setting. Thus, they proposed a taxonomy to accommodate some of such situations and interrelate time/space matrix with information sharing, communication and coordination. In the same year, Elmarzouqi *et al.* [51] approached the Augmented Continuum of Collaboration Model (ACCM), which is also focused in the CSCW characteristics (collaboration, cooperation, and coordination), and relating them with ACCM components: co-production, communication, and conversation. In summary, this taxonomy is based on the 3C model, with addition of conversation, regulation and awareness. The Mittleman *et al.*'s [8] taxonomic study includes nine architectural implementations, with a specific granularity, to classify the attributes of groupware systems. This taxonomic effort follows a previous work [55] based on the encapsulation of collaboration patterns to classify collaborative work.

A conceptual framework was suggested for the design of groupware user interfaces [52] with a focus on shared context, visualization area, activity, division of labor, task types, geographical information, people, events, time interval, object, strategy, as well as rules. Following the proliferation of systems and algorithms in industry and academy, it was proposed an evaluation model for collaboration systems at an information seeking perspective [53] with the domains of intent, depth, concurrency, and location. Furthermore, Briggs *et al.* [54] give a social-technical perspective to define "seven areas of concern for designers of collaboration support systems", subdividing them by goals, products, activities, patterns, techniques, tools, and scripts. This classification model represents the starting point for the classification model presented here.

5   Socio-technical Requirements to Support Collaboration

Taxonomic proposals of CSCW and groupware are varied, with a fair amount of differences. Penichet *et al.* [10] argued that one of the main reasons to this variety relies on the increased complexity of groupware tools. An overarching classification model is proposed to categorize collaboration requirements for a more social-oriented groupware development. According to Johnson [56], the creation and refinement of classification systems and taxonomies are crucial processes in theory development, where the categories of a classification model should be mutually exclusive, exhaustive, and logically interrelated. The socio-technical classification model proposed here attempts to bring a *continuum* of collaboration dimensions, which problem relies on the lack of standardization of categories proposed in literature without terminological consensus. In the Figure 2 are shown socio-technical requirements for collaboration, organizing a set of categories taken from literature. The model intends to tackle the evident lack of consensus concerning to the conceptual structure of cooperative work and groupware at a combined perspective, comprising technical requirements and work dimensions in an unified classification model. The taxonomic elements of the scheme presented here are fully based in CSCW and group generic literature, which was extracted taking into account their temporal persistence, bibliometric impact, complementarity, and logical consistence. The "blocks" and "meta-blocks" of this model establish a set of domains according to their granularity, being structured at an hierarchical way.

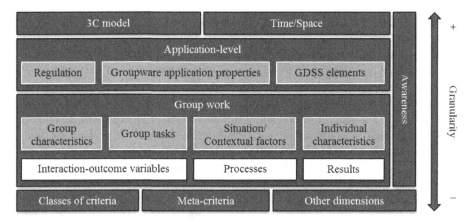

Fig. 2. Socio-technical classification model of CSCW and groupware

The first category of this literature-based classification model is the *3C model*. This category can be systematized into an interactive cycle through the well-known modes of collaboration. *Communication* can be understood as an interaction process between people [17], involving explicit or implicit information exchange, in a private or public channel. The users' messages can be identified or anonymous, and conversation may occur with no support, structured or intellectual process support, with associated protocols. As a requirement, groupware needs to be able to support the conversation between two or more individuals, in an one-to-one, one-to-many or many-to-many setting. *Coordination* was defined by Malone & Crowston [33] as management of interdependencies between activities performed by multiple actors, which are based on the mutual objects that are exchanged between activities (e.g., design elements, manufactured parts, or resources). Some categories related to the coordination in the literature are: planning, control models, task/subtask relationship and information management, mutual adjustment, standardization, coordination protocol, modes of operation, and so on. In order to effectively support coordination, groupware needs to fulfill three important requirements: time management, resources, or shared artifacts produced along the activity chain. *Cooperative work* arrangements appear and dissolve again. Oppositely to conflict [17], cooperation occurs when a group works toward a common goal [33] with high degrees of task interdependencies, sharing the available information by some kind of shared space [6]. Cooperation categories can range from production (co-authoring), storage or manipulation of an artifact, to concurrency, access or floor control. Technologically, cooperation is supported by systems with capabilities to send or receive messages, synchronously and/or asynchronously [8, 35], and also develop or share documents [32], which are identified as requirements in socio-technical classification model. Furthermore, this terminology was adopted by representing a predominant view in CSCW field, although not totally agreed by some researchers.

Collaboration can occurs in a specific time (synchronous and/or asynchronous) and place (co-located or remote), and may have high or low levels of predictability. If we granulize the *Time/Space* category, a set of subdomains can be distilled, more precisely: session persistence, delay between audio/video channels, reciprocity and homogeneity of channels, delay of the message sent, and spontaneity of collaboration. Complementary, it can be useful to define contextual issues to improve work dynamics.

In order to cooperate in the current polymorphic settings, group members must be aware of other's activities, creating group *awareness* at the workspace. The collaboration cycle is bounded by awareness, which is the perception of group about what each member develops, and the contextual knowledge that they have about what is happening within the group [8]. In this sense, awareness mechanisms are essential in collaboration systems to reduce work losses. It characterizes space and atmosphere, activity, object, human, and meta-dimensions such as presence, influence, and abilities.

The *application-level* category identifies a set of typologies for groupware systems. Mittleman *et al.* [8] proposed several categories and subcategories to classify collaboration technology according to its focus on the group level, covering work over a period of time: i) jointly authored pages (conversation tools, polling tools, group dynamics, and shared editors); ii) streaming technologies (desktop/application sharing, audio conferencing, and video conferencing); iii) information access tools (shared file repo-sitorics, social tagging systems, search engines, and syndication tools); and iv) aggregated systems. Complementary, a large set of meta-domains can be extensively identified (e.g., message systems, information sharing technologies, GDSS, project, virtual workspaces, meeting minutes/records and electronic meeting rooms, process or event management systems, chat/instant messaging, notification systems, group calendars, collaboration laboratories, bulletin boards, data mining tools, e-mail, workflow systems, intelligent agents, and so on. In this sense, a lack of consensus about groupware domains is a serious challenge to overcome. As a subcategory of groupware systems, *regulation* means the representation of mechanisms that enable participants to organize themselves into a shared environment, where the regulation of collaboration activities concerns the definition and evolution of work rules to ensure conformity between the activity and group goals [38]. It is worth noting that some of the regulation dimensions achieved from literature are: arenas (location); actors (roles, places, and positi-ons); tools (regulative or not); roles (thematic or causal); rules (constraints, norms, or work rules); types of interaction; interactive scenarios; and objects (means of communication and product of collaboration). The *groupware application properties* can be constituted by functional properties of collaboration tools: architecture, functional and quality properties, group processes support, collaboration interface (portal, devices, or physical workspace), relationships (collection, list, tree, and graph), core functionality, content (text, links, graphic, or data-stream), supported actions (receive, add, associate, edit, move, delete, or judge), identifiability, access controls, alert mechanisms, intelligent/semi-intelligent software components, awareness indicators, and platform. *GDSS elements* can include hardware, software, organizationware and people support.

Groups can be defined as "social aggregations of individuals" with awareness of its presence, conducted by its own norms, and supported by task interdependencies towards a common goal in a shared purpose or work context [46]. In this sense, a group is constituted by particular characteristics, such as: size (3 to 7, >7), composition, location, proximity, structure (leadership and hierarchy), formation, group awareness (low or high, and cohesiveness), behavior (cooperative or competitive), autonomy, subject, and trust. The *group members* have a personal background (work experience, training, and educational), skills, motivation, attitude towards technology, previous experience, satisfaction, knowledge, and personality. According to McGrath [17], *group tasks* can be subdivided in creativity, planning, intellective, decision-making (choosing, evaluation and analysis, search, report, and survey), cognitive-conflict, mixed-motive, contests/battles/competitive and performances/psychomotor, having a specific complexity associated to each task. The subcategories can be supported by cultural impact, goals, interdependency or information exchange needs, bottlenecks, or process gain and loss. The *contextual* or *situational factors* can range from organizational support (rewards, budget, and training), cultural contexts (trust or equity), physical setting, environment (competition, uncertainly, time pressure, and evaluative tone), and business domain at an organizational way. Interaction variables are related to group factors: i) *interaction outcome variables*, such as group outcomes (quality of group performance, collaboration processes, and group development), individual outcomes (expectations and satisfaction on system use, appreciation of group membership, and individual breakdowns in system use), and system outcomes (enhancements and affordances); ii) *processes,* including individual, interpretation, motivation and performance dimensions; and iii) *results*, specifically individual rewards, group vitality, and organizational results).

The independent variables are focused on *classes of criteria* (functional, technical, usability, and ergonomics), *meta-criteria* (scalability, and orthogonality) and *complementary dimensions* without a specific domain. Some of the other dimensions that can characterize a socio-technical collaboration scenario are: work coupling, shared tasks and goals, information richness and type, control centralization, activities, division of labor, patterns, techniques, scripts, assistance, learning monitoring, interaction degree, assertion, events, strategy, social connectivity, content management, process integration, sharing (view/opinion, knowledge/information, and work/operation), protection, distributed processes loss, or depth of mediation. However, categories such as accessibility are partially forgotten from CSCW literature studies.

6   Conclusions and Future Work

This paper concludes that there are, in fact, common dimensions to the several taxonomies addressed by literature with more or less regularity. The growth of importance of CSCW characteristics, in opposition to categories such as group issues, might suggest a relation between them. The 3C model is well accepted, widely used, and extremely useful to classify collaboration settings. A literature review leads us to establish requirements solid enough to characterize existing tools or group work dynamics, and the development of new ones. Some taxonomies (e.g., [5, 17]) have an high number of citations at a bibliometric level, which reflects the importance of organizational and group work dynamics, application-level, 3C model, and time/space. Oppositely, it is approached the studies of Okada [50] and Giraldo *et al.* [52] without citations.

A collaboration project can fail due to the lack of literature-based and/or empirical studies about requirements and satisfactory practices. Theories, principles or conceptual frameworks provided by psychology, sociology or anthropology in work contexts can improve group work [46]. The socio-technical classification model can reproduce a mutual adaptation process between groupware and people observable from a collaboration perspective. A key contribution of this paper is to provide a systematic review of how groupware and work settings have been presented in CSCW literature. In this context, this work relies on a community-centered building model with literature support to understand the complex taxonomic phenomena, expanding and adapting existing approaches to create taxonomies and design a contemporary classification model. Organizations, academies, industries and other social entities can apply this model to retrieve insights about collaboration dynamics and improve the quality of interaction.

In future, the literature classification can be done under a more detailed framework, where the taxonomic dimensions would be subdivided in several features. This aspect could help to find the functional requirements of groupware tools, making a step forward towards a more wide and consensual taxonomy of CSCW and groupware. There is a need to translate these theoretical propositions into practical guidelines using a set of methods such as semi-structured interviews or ethnography to find evidences about current virtualized work practices, breaking the 'tunnel effect' of CSCW terminology.

References

1. Baladi, M., Vitali, H., Fadel, G., Summers, J., Duchowski, A.: A taxonomy for the design and evaluation of networked virtual environments: its application to collaborative design. International Journal on Interactive Design and Manufacturing 2(1), 17–32 (2008)
2. Grudin, J., Poltrock, S.E.: Taxonomy and theory in Computer Supported Cooperative Work. In: Kozlowski, S.W.J. (ed.): The Oxford Handbook of Industrial and Organizational Psychology. Oxford University Press, New York (in press)
3. Ackerman, M.S.: The intellectual challenge of CSCW: the gap between social requirements and technical feasibility. Human-Computer Interaction 15(2), 179–203 (2000)
4. Schmidt, K., Bannon, L.: Taking CSCW seriously: supporting articulation work. Computer Supported Cooperative Work 1(1), 7–40 (1992)
5. Ellis, C.A., Gibbs, S.J., Rein, G.L.: Groupware: some issues and experiences. Communications of the ACM 34(1), 38–58 (1991)
6. Grudin, J.: Computer Supported Cooperative Work: History and focus. Computer 27(5), 19–26 (1994)
7. Weiseth, P.E., Munkvold, B.E., Tvedte, B., Larsen, S.: The wheel of collaboration tools: A typology for analysis within a holistic framework. In: Proceedings of the ACM Conference on Computer-Supported Cooperative Work (CSCW 2006), pp. 239–248 (2006)
8. Mittleman, D.D., Briggs, R.O., Murphy, J., Davis, A.: Toward a Taxonomy of Groupware Technologies. In: Briggs, R.O., Antunes, P., de Vreede, G.-J., Read, A.S. (eds.) CRIWG 2008. LNCS, vol. 5411, pp. 305–317. Springer, Heidelberg (2008)
9. Verginadis, Y., Papageorgiou, N., Apostolou, D., Mentzas, G.: A review of patterns in collaborative work. In: Proceedings of the 16th ACM International Conference on Supporting Group Work (GROUP 2010), pp. 283–292. ACM, NY (2010)
10. Penichet, V.M.R., Marin, I., Gallud, J.A., Lozano, M.D., Tesoriero, R.: A classification method for CSCW systems. Electronic Notes in Theoretical Computer Science 168, 237–247 (2007)
11. Schmidt, K.: Analysis of cooperative work – a conceptual framework. Risø National Laboratory, DK-4000 Roskilde, Denmark (Risø-M-2890) (June 1990)
12. Kitchenham, B., Brereton, O.P., Budgen, D., Turner, M., Bailey, J., Linkman, S.: Systematic literature reviews in software engineering – a systematic literature review. Information and Software Technology 51, 7–15 (2009)
13. Webster, J., Watson, R.T.: Analyzing the past to prepare for the future: writing a literature review. MIS Quarterly 26(2), 13–23 (2002)
14. Thomas, D.R.: A general inductive approach for qualitative data analysis. University of Auckland, New Zealand School of Population Health (2004)
15. Gregor, S.: The nature of theory in Information Systems. MIS Quarterly 30(3), 611–642 (2006)
16. Bafoutsou, G., Mentzas, G.: Review and functional classification of collaborative systems. International Journal of Information Management 22, 281–305 (2002)
17. McGrath, J.E.: Groups: interaction and performance. Prentice-Hall, Englewood Cliffs (1984)
18. Mintzberg, H.: A typology of organizational structure. In: Miller, D., Friesen, P.H. (eds.) Organizations: A Quantum View, pp. 68-86. Prentice-Hall, Englewood Cliffs (1993); Reprinted in Baecker, R. (ed.): Readings in groupware and Computer-Supported Cooperative Work, pp. 68–86. Morgan Kaufmann, San Mateo (1993)
19. Zigurs, I., Munkvold, B.E.: Collaboration technologies, tasks, and contexts: evolution and opportunity. In: Galletta, D., Zhang, P. (eds.) Human-Computer Interaction and Management Information Systems: Applications, vol. II, pp. 143–169. M. E. Sharpe, Armonk (2006)

20. Mikki, S.: Google Scholar compared to Web of Science: A literature review. Nordic Journal of Information Literacy in Higher Education 1(1), 41–51 (2009)
21. Carter, L.F., Haythorn, W.W., Howell, M.A.: A further investigation of the criteria of leadership. Journal of Abnormal and Social Psychology 46(6), 589–595 (1950)
22. Shaw, M.E.: Some effects of problem complexity upon problem solution efficiency in different communication nets. Journal of Experimental Psychology 48(3), 211–217 (1954)
23. McGrath, J.E., Altman, I.: Small group research: a synthesis and critique of the field. Holt, Rinehart & Winston, Chicago (1966)
24. Shaw, M.E.: Scaling group tasks: a method for dimensional analysis. JSAS Catalog of Selected Documents in Psychology 3, 8 (1973)
25. Bui, T., Jarke, M.: Communication requirements for Group Decision Support Systems. Journal of Management Information Systems 11(4), 9–20 (1986)
26. DeSanctis, G., Gallupe, R.B.: A foundation for the study of Group Decision Support Systems. Management Science 33(5), 589–609 (1987)
27. Jelassi, M.T., Beauclair, R.A.: An integrated framework for Group Decision Support Systems design. Information and Management 13(3), 143–153 (1987)
28. Nunamaker Jr., J.F., Briggs, R.O., Mittleman, D.: Electronic meetings to support group work. Communications of the ACM 34(7), 40–61 (1991)
29. Johansen, R.: Groupware: computer support for business teams. The Free Press, New York and London (1988)
30. Borghoff, U.M., Schlichter, J.H.: Computer-Supported Cooperative Work: introduction to distributed applications. Springer, USA (2000)
31. Ellis, C.A., Wainer, J.: A conceptual model of groupware. In: Proceedings of the ACM Conference on Computer Supported Cooperative Work (CSCW 1994), pp. 79–88 (1994)
32. Grudin, J., Poltrock, S.E.: Computer-Supported Cooperative Work and groupware. In: Zelkowitz, M. (ed.) Advances in Computers, vol. 45, pp. 269–320. Academic Press, Orlando (1997)
33. Malone, T.W., Crowston, K.: The interdisciplinary study of coordination. ACM Computing Surveys 26(1), 87–119 (1994)
34. Jarczyk, A., Loffler, P., Volksen, G.: Computer Supported Cooperative Work (CSCW) – state of the art and suggestions for the future work. Internal Report, Version 1.0, Siemens AG, Corporate Research (1992)
35. Mentzas, G.: Coordination of joint tasks in organizational processes. Journal of Information Technology 8, 139–150 (1993)
36. Ellis, C.A.: An evaluation framework for collaborative systems. Technical Report, CU-CS-901-00, Department of Computer Science, University of Colorado at Boulder, USA (2000)
37. Fjermestad, J., Hiltz, S.R.: An assessment of Group Support Systems experimental research: methodology and results. Journal of Management Information Systems 15(3), 7–149 (1998)
38. Ferraris, C., Martel, C.: Regulation in groupware: the example of a collaborative drawing tool for young children. In: Proceedings of 6th International Workshop on Groupware (CRIWG 2000), pp. 119–127 (1998)
39. Schümmer, T., Lukosch, S.: Patterns for computer-mediated interaction. John Wiley & Sons Ltd., Chichester (2007)
40. Sarma, A., Van Der Hoek, A., Cheng, L.T.: A need-based collaboration classification framework. In: Proceedings of the ACM Conference on Computer Supported Cooperative Work (CSCW 2004), Chicago, II., RC 23339 (2004)

41. Boughzala, I., Assar, S., Romano Jr., N.C.: An E-government field study of process virtualization modeling. In: de Vreede, G.J. (ed.) Group Decision and Negotiation, vol. 154 (2010)
42. Stefik, M., Foster, G., Bobrow, D.G., Kahn, K., Lanning, S., Suchman, L.: Beyond the chalkboard: computer support for collaboration and problem solving in meetings. Communications of the ACM 30(1), 32–47 (1987)
43. Kraemer, K., King, J.: Computer-based systems for cooperative work and group decision making. ACM Computing Surveys 20, 115–146 (1988)
44. McGrath, J.E., Hollingshead, A.B.: Groups interacting with technology: ideas, evidence, issues, and an agenda. Sage Publications, Thousand Oaks (1994)
45. Coleman, D.: Groupware: technology and applications. Prentice Hall, Upper Saddle River (1995)
46. Pumareja, D.T., Sikkel, K.: An evolutionary approach to groupware implementation: the context of requirements engineering in the socio-technical frame. Technical Report TR-CTIT-02-30, Centre for Telematics and Informational Technology, University of Twente, Enschede (August 2002)
47. Andriessen, J.H.: Working with groupware: understanding and evaluating collaboration technology. Springer, London (2002)
48. Bolstad, C.A., Endsley, M.R.: Tools for supporting team collaboration. In: Proceedings of the 47th Annual Meeting of the Human Factors and Ergonomics Society, pp. 374–378. HFES, Santa Monica (2003)
49. Neale, D.C., Carroll, J.M., Rosson, M.B.: Evaluating Computer-Supported Cooperative Work: models and frameworks. In: Proceedings of the ACM Conference on Computer Supported Cooperative Work (CSCW 2004), pp. 112–121 (2004)
50. Okada, K.: Collaboration support in the information sharing space. IPSJ Magazine 48(2), 123–125 (2007)
51. Elmarzouqi, N., Garcia, E., Lapayre, J.C.: ACCM: a new architecture model for CSCW. In: Proceedings of the 11th International Conference on Computer Supported Cooperative Work in Design (CSCWD 2007), Melbourne, Australia (2007)
52. Giraldo, W.J., Molina, A.I., Gallardo, J., Collazos, C.A., Ortega, M., Redondo, M.A.: Classification of CSCW proposals based on a taxonomy. In: Proceedings of the 13th International Conference on Computer Supported Cooperative Work in Design (CSCWD 2009), pp. 119–124. IEEE Computer Society, Washington, D.C. (2009)
53. Golovchinsky, G., Pickens, J., Back, M.: A taxonomy of collaboration in online information seeking. Paper presented at the Collaborative Information Retrieval Workshop at the Joint Conference on Digital Libraries (JCDL 2008), pp. 1–3 (2008)
54. Briggs, R.O., Kolfschoten, G., Vreede, G.-J., Albrecht, C., Dean, D.R., Lukosch, S.: A seven-layer model of collaboration: Separation of concerns for designers of collaboration systems. In: Proceedings of the Thirteenth International Conference on Information Systems, Phoenix, AZ (2009)
55. Briggs, R.O., De Vreede, G., Nnunamaker Jr., J.F.: Collaboration engineering with think Lets to pursue sustained success with Group Support Systems. Journal of Management Information Systems 19(4), 31–64 (2003)
56. Johnson, D.P.: Contemporary sociological theory: an integrated multilevel approach. Springer, New York (2008)

Normal Users Cooperating on Process Models: Is It Possible at All?

Alexander Nolte and Michael Prilla

Department Information and Technology Management, Institute for Applied Work Science,
University of Bochum, Germany
{alexander.nolte,michael.prilla}@iaw.rub.de

Abstract. Can normal people use process models for self-directed cooperation, that is, without expert guidance? According to modeling experts and corresponding contemporary research, they cannot, because they lack competencies for such usage. While the importance of artifacts such as texts, pictures and diagrams to cooperative work has been shown in many studies in CSCW and related fields, there are no answers to this question from our discipline. This paper aims at exploring this contradictory situation by exploring how users without or with little modeling practice work with models. Based on an exploratory study, we show opportunities and barriers to self-directed cooperative work with models and derive requirements for tool support. These results are compared with existing work and show that despite the special characteristics of process models, patterns known from the usage of other artifacts can also be observed in cooperative work with models. Users also showed behavior typically attributed to modeling experts, thus transcending such generic cooperation tasks.

Keywords: Cooperation support, process models, lay modeling.

1 Introduction: Can Models Be Used to Support Cooperation?

Process models are common tools in organizations. They are used for describing, designing, analyzing and improving work or business processes and thus describe work done by many people in organizations. It seems reasonable to use them collaboratively involving all relevant people to integrate their views: This helps to make process models represent real world processes instead of idealized views [1], provides a basis for negotiating and supporting processes in which people work together [2] and supports users when they want to know how certain parts of work are done [3]. Despite these potentials, models are currently only used by a small group of experts [3].

Due to the aforementioned potentials, it is desirable to enable all people to collaboratively work with or on process models on their own [3, 4]. This shows to be especially difficult for the group of people, who have not been trained to use models, which we refer to in this paper as lay users or non-expert modelers. Current cooperation research does not cover lay users' work with models and thus, there are no insights into support for this. In addition, many modeling experts deny lay users the capability to use models because of a number of reasons (e.g. [1, 5]) that seem to be

V. Herskovic et al. (Eds.): CRIWG 2012, LNCS 7493, pp. 57–72, 2012.

both obvious and intuitive: Models are more complex to use and less common than other types of content such as e.g. text. In addition, modeling languages are often very complex to use. Furthermore, only few people in organizations are familiar with modeling languages and tools. But does this mean that models do not qualify as proper artifacts of cooperative work for lay users? Based on experiences with using process models (e.g. [6]), in this paper we do not incline to this view and state that lay users can use models cooperatively if we can find ways to support them properly. Thus, the research questions for the work presented in this paper are:

RQ1: **Under which conditions can lay users use process models as artifacts in their cooperation?** Our work focuses on minimal support for cooperative usage of process models in order to derive basic conditions for this usage.

RQ2: **How can cooperative usage of process models by lay users be supported?** Depending on the answers to the former question, this follow-up question asks for social and technical support demands for lay user in using process models.

As stated above, to the knowledge of the authors there is no work available explicitly dealing with cooperative work of lay users on or with process models – on the contrary, many experts suggest that this is not possible without expert guidance anyway. The work described in this paper strives to shed light on this, looking for answers to the questions stated above. Being part of ongoing work [4, 7–9], this paper focuses especially on cooperative usage of models (instead of e.g. cooperative model creation). The paper starts with reviewing relevant literature (section 2). Afterwards we describe a study (section 3) in which participants cooperated on models. Observations from the study (section 4) and their analysis (section 5) indicate that people can use models in self-regulated cooperation surprisingly well. The paper continues with a description of prototypes for lay user model interaction (section 6) and concludes with a summary and future work (section 7).

2   Process Models and Cooperative Work

Investigating the aforementioned research questions requires reviewing the relevant literature on process models, cooperative modeling and the cooperative usage of process models. This section gives an overview of respective work and summarizes it into dilemmas and hopes for cooperation of lay users on process models.

2.1   What Makes Process Models Special?

This paper focuses on the usage of process models by people who have not been trained to use them. In order to understand why this is a difficult task and why most experts doubt that such non-expert modelers can use process models, one needs to understand what we mean by *process models* and work with them.

By *model*, we refer to a graphical representation of real world phenomena that follows a certain notation, including syntactical elements of a modeling language, and expressing certain semantics of the real world. A process model is a specialization of this, which represents sequences of tasks including predecessors, ramifications and logical gateways in processes. Thus, knowing a notation seems to be an indispensable skill when working with models.

A process model – just like any model – represents certain details of real work phenomena by deliberately providing a level of abstraction that modelers perceive as suitable for the purpose of the model – a process model for software development will contain many details on technical aspects, while a model for organizational improvement might focus on detailed steps of work. Thus, a process model reduces the complexity of real world processes into the representational details of the model. The abstraction used for this often makes it hard for non-expert-modelers to use process models. Moreover, a process model is the result of mapping (individual) mental models of processes to a modeling notation [10]. As people have different mental models of processes they are involved in [10] and necessarily possess different information on process details (e.g. [2]), process models are dependent on the modeler – the representation can only include what is known at modeling time and will represent the mental models of people involved in the modeling process to some extent. Moreover, people will have different perspectives on processes and will know different details on process parts, depending on the way they are involved in the process. Therefore, process modeling and cooperating on processes always means integrating perspectives. This complexity also explains why models are increasingly perceived as part of cooperative activities (e.g. [8, 11, 12]).

The description above shows how process models differ from other virtual artifacts used in cooperation. In contrast to textual documents such as papers or checklists, they use notations most people are not familiar with and in contrast to other graphical models such as organograms, process models are highly subjective and depending of perspectives integrated into them. As described above, this makes it hard for normal users to interact with process models and poses a huge challenge for research.

2.2 Cooperative Modeling of Processes

The modeling of processes is a complex task. Especially if different perspectives upon the process have to be negotiated and represented within these models, process participants have to be integrated into model development. Participation concepts range from experts creating process models based upon upfront interviews [13] to directly involving stakeholders into model development within co-located modeling workshops [11]. Within these workshops stakeholders verbally contribute their view of the process (c.f. [6, 14]). They are supported by a number of experts, which **structure** the communication process, **translate** contributions by the participants into model elements and **operate** the modeling tool thus altering the model according to the contributions. These tasks are often distributed between different experts that run the modeling workshop. The number of experts fulfilling these tasks ranges from two (facilitator and modeler; c.f. [6]) up to five (facilitator, modeler, process coach, recorder, gatekeeper; c.f. [15]) depending on the exact procedure and goal of the modeling.

The common procedure of modeling workshops is that a facilitator asks the participants about when a task is performed, who performs it and which tools or information is required. By asking which action follows the previous one the participants walk through the process thus ensuring that no action will be missed [6]. Limiting participation to verbal contributions comes due to the belief that non-expert modelers cannot express themselves in a modeling notation because of their complexity [16].

There is a significant body of research upon workshop participants directly contributing to model development. Within these approaches – named collaborative modeling [17] – the workshop participants operate the modeling tool themselves thus also taking over the task of translating their contributions into elements according to the modeling notation. These approaches however only work for people who use models regularly [18] or people who are learning a modeling notation [19].

2.3   Cooperative Usage of Process Models

Models and their usage by have been a topic CSCW research, but there is only little work available on actual cooperative usage of process models. Early work has regarded models suspiciously, as they were perceived to be idealized and thus to not inform design groupware properly (e.g. [1]). Consequently, this work has created methods of collaborative model usage (see section 2.2). More recent research shows that – sensibly used – process models can support people in making perspectives explicit, exchanging them and negotiating a common understanding [20] as well as that models are artifacts supporting cooperation among professionals [21]. However, similar to approaches of participatory modeling (see section 2.2), these insights are based on model usage supported by experts. Thus, while it indicates the potential value of models in cooperation, existing work falls short in describing the extent to which people can use models cooperatively on their own and without additional guidance.

More general, there is a lot of work on the role of virtual artifacts in cooperative work such as textual documents [22] or pictures [23]. This work shows that virtual artifacts in general can support cooperation, e.g. by using them as plans for action [24] describing individual and cooperative work and as "protocols" for coordinating work [1], which describe rules for cooperation. Artifacts may support coordination of work [2] and can be "means for transferring tacit knowledge about action affordances among people" [25]. For pictures, it has been shown that artifacts can guide communication and support or alternatively express communicative statements [23].

There is no similar work for models, although they represent processes that are mostly conducted collaboratively and contain knowledge on individual and collaborative work. This raises the question whether the potentials of other virtual artifacts as described above can also be tapped by cooperative usage of models. The work described above shows that if this is the case, they must have a sustainable and nontrivial impact on people and their communicative interaction (e.g. guiding their communication as described above), which then in turn shapes people's examination and adoption of these artifacts (e.g. using them for knowledge transfer).

2.4   Models in Cooperation – Dilemmas and Hopes from Practice

Given the characteristics of models (section 2.1) and the lack of research on collaborative model usage without expert guidance (sections 2.2 and 2.3), we may doubt whether non-expert-modelers can use process models on their own: Research leaves this question unanswered but describes multiple dilemmas to be overcome:

- To use process models cooperatively and without expert guidance, people must understand them and to make sense of them, but it is doubted that they can do this without support because of the inherent complexity of process models and the need to use modeling notations (e.g. [5] and section 2.1).

- Stakeholder participation in modeling is currently supported for co-located modeling workshops with expert facilitation and guidance, but this results in the 'facilitator bottleneck' described above, making cooperative process modeling an infrequent and hard to organize task [4].
- Using models in cooperation requires interaction. This needs using modeling tools, which are difficult to use for people who have not been trained to do so [3, 4].

Given these dilemmas, one may ask whether it is worthwhile to investigate models as artifacts supporting cooperation. However, we know that models are commonly used in practice [26] and in our previous work we experienced that non-expert-modelers can understand and use them properly [6]. In particular, we found that (sometimes)

- people can use models to *support cooperation* with minimal guidance [3],
- models *support communication* between modeling and domain experts [3],
- people can make use of models and *understand* them [8],
- people can *contribute to models* given adequate means to do so [4].

It must be noted, however, that these findings indicate the general feasibility of using process models in cooperation, but do not make any assumptions on concrete tasks and qualities of cooperative work. However, they strongly suggest that models can be valuable artifacts of cooperation if we understand how they are used.

3   Study: Exploring Cooperative Usage of Process Models

In order to observe the impact of models on cooperation, we conducted a study to explore whether models can be used in the same way as other artifacts, thus leaving traces in communication and cooperation of people using them. This section gives an overview of the setup, setting and evaluation methodology of the study.

3.1   Experimental Setting and Task

In our study, participants worked on models of processes they were familiar with. This included software development and library management and usage (c.f. Table 1). We conducted five workshops with two participants each, lasting about 30-45 minutes. Three participants considered themselves to be modeling experts, two use modeling tools occasionally and five were new to modeling.

Table 1. Scenarios used in the experiment, including pairs (Px) and participants (Px.1/2)

Scenario	Pairs
(1) Bug fixing in software development	P1 (P1.1, P1.2), P3 (P3.1, P3.2), P4 (P4.1, P4.2)
(2) Book ordering in a library	P2 (P2.1, P2.2), P5 (P5.1, P5.2)

In our study we used the SeeMe modeling language as it has been proven to be fairly comprehensible for lay users [6]. It is also pretty similar in terms of elements used to other popular methods such as BPMN and EPCs. So our results can – up to a certain extent – be transferred to scenarios in which other languages are used.

In our study, we used an interface that allows for people to directly interact with models on an interactive large screen (c.f. Fig. 1 right). It enables aligning elements, changing element types and marking elements by touch input.

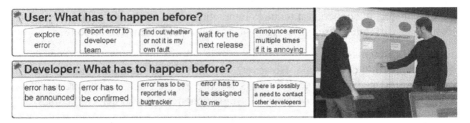

Fig. 1. Study participants comparing their respective view of a process (left) and altering the process model using an interactive large screen (right)

In preparation for each workshop we asked the participants to create a model of the process using a text-based interface [4]. In study the participants sorted the sequence of their own models in solitude. This resulted in two models per pair, each representing the perspective of one participant (Fig. 1 left). In the following collaborative interaction phase, the participants were asked to identify and discuss differences between these two models and to these differences directly in the model, using the interactive screen (Fig. 1 right). During the whole experiment, a facilitator was present explaining the tasks to the participants. He did not intervene into model related tasks.

3.2   Methodology

The study described here is exploratory and aims at observing non-experts collaborating on process models. The interaction of the pairs with the models was videotaped by cameras and microphones installed in the room, including a screen capture of the modeling tool. For analysis, we watched the videos repeatedly and noted observations, including the time of their occurrence, a description of the sequence we had seen and our interpretation. In an approach similar to grounded theory coding (cf. [27]), we refined the results of coding by synergizing codes from single observations into a common scheme of observations. To narrow down this scheme and focus on models in cooperation, we extracted categories related to cooperation. This selection contains 150 codes, which were refined into 32 comprehensive codes and four categories. We used the four categories to structure the following section (headlines).

4   Observations from the Study

The assumption underlying our study is that if models can be used in cooperative work of non-expert-modelers, we can observe non-trivial impacts of the process models on communication between people and on the way they cooperate (section 2.3). This section describes our observations with a special focus on process models as guides for cooperative work, process models as part of the articulation during cooperation and process modeling as part of cooperative model usage.

4.1   Process Models as Guides in Cooperative Work

In many situations, our participants were led by the process models in their cooperative interaction. While it sounds trivial that people are led by the artifact they are using, we found that the participants engaged in interaction related to the characteristics of process models (see section 2.1) and that they went into discussing the process models in a level of detail that we did not ask for explicitly in the study introduction.

Understanding: Discursive Explanation of Sequences and Structures
All pairs explained the meaning of model elements to each other. This was often triggered by questions such as "What does this [pointing to element][1] have to do with bug processing?" (P3.2). While this could be expected beforehand, we also found more complex interaction, e.g. when they compared elements from different models. (Fig. 2 left) shows a characteristic occasion of this, in which P2.2 asks her partner whether two elements from their processes would mean the same, pointing to both elements. Such observations suggest that it was the actual structure of the process model that lead the interaction and not the sheer presence of an arbitrary shared artifact.

Fig. 2. Different usage of references to multiple model elements (P2): Comparing models and perspectives by pointing to different elements of different models (left) and telling a story while pointing to a range of elements in the same model (right)

Structuring Communication: Using Models for Storytelling
Without asked to do so, three participants started to tell stories about their respective model to explain it to their partners. In Fig. 2 (right), P2.2 is telling her partner a story about the process part between her right and left arms: "Once I know for which purpose I need the book [first element] and when I know whether to buy it or not [second element], then I send you an email [third element] and (…)" (P2.2). This observation suggests that the participants perceived the models as personal expressions of their individual view on the process rather than an arbitrary graphical artifact.

Contextualizing Communication: Using Process Models for Orientation
Four pairs also used the process models to contextualize their communication: "First, you look for the reason [pointing to element] and when you find the reason, you think of an idea for solving the problem [pointing to element]. Then you check and implement the idea [pointing to two elements] (…)" (P1.1). This can also be seen in Fig. 3, in which P2.2 points to different elements of her process model to show P2.1 what she is talking about. Again, this shows how the participants made use of model semantics (i.e. their sequence) and thereby integrated the process models into their conversation.

---

[1] Explanations in brackets were added in the quotes of participants for better understanding.

Fig. 3. P2.1 using gestures to model elements to depict what she is talking about, starting in the first half and going to the end of the model (sequence captured within 6 seconds)

From the deictic and verbal references to the models, it can be seen that the participants did not talk about arbitrary drawings, but made sense of the models. The models became artifacts guiding and inspiring their conversation even beyond tasks of the study. This exemplifies how models can be used to support and drive cooperation.

4.2 Process Models as Part of the Articulation during Cooperation

If models support cooperation (like other artifacts), there are traces of them in the communication of participants (see section 2.3). In the study, we observed such traces, which suggest that the process structure of the models influenced communication.

References to Models as Support and Elements of Communication
Two pairs used model elements as part of their communication as in "We also need to take this [pointing to element] into account" (P3.2). Model parts were also used as statements, e.g. when P1.1 was asked "Don't you need to implement the solution first?" and answered by pointing to an element, saying "That happens here!". Bearing in mind that neither P1 nor P3 included a modeling expert, these observations show a deep integration of modeling language into their conversation, which suggests that discursive interaction with models is feasible for many non-expert-modelers.

Developing a Common Understanding
All pairs engaged in processes of grounding [4] when working in the models. This can be seen by statements such as "Ah, now I understand what you mean [pointing to an element]" (P1.2) and "and the difference in your model is (…)?", which was answered by P3.2 with "Yes, exactly". This shows that the participants understood the semantics of their models, accepted them as representations of their mutual perspectives on the process and based their communication on them.

Getting Aware of and Transferring Knowledge from Other Perspectives
Four of five pairs explicitly tried to understand details behind each other's perspective. This can be seen in statements such as "I have to admit that I have not thought about what [you do] to be prepared for my orders" (P5.1 telling P5.2 about his knowledge on her work). Pairs also switched perspectives in order to understand each other better: "I know that you are doing this, but I did not have it in mind" (P5.1 reflecting on P5.2's work). This suggests that the participants related meaning to the process models and shows how they used it as a medium for knowledge exchange.

These observations indicate that the participants integrated process model semantics into their communication and that they also looked deeper into the models, investigating details and asking their partners about them. This is non-trivial interaction and shows how non-expert-modelers can be supported in articulation and knowledge transfer on processes by corresponding process models.

4.3   Process Modeling as Part of Cooperative Model Usage

As described in section 2.2, most cooperative activities of non-expert-modelers rely on expert facilitators, as normal users are not expected to have the capabilities to perform them. In our study, however, we observed several modeling activities in which the participants actually performed activities performed by experts otherwise.

Asking for Feedback on Manipulations and Negotiating Process Content
For four of the pairs, we observed one partner explicitly asking for feedback when she made changes in the model, e.g. "What do you think? Correct?" (P1.2 after a change). Fig. 4 right shows an example of this. In addition, all pairs negotiated common perspectives or sequences in the processes. This included negotiations on contrary positions, e.g. when P5.2 stated "To me this only makes sense if (...)" and P5.1 replied "I always want to have this!". Fig. 4 (left) shows P5.1 and P5.2 arguing about this. The participants felt responsible to create a common understanding and represent it in their models. These are activities typically delegated to a facilitator (see section 2.2).

Fig. 4. Discussion of contrary positions on model elements during work in the experiments (P5, left) and explicit feedback request after changing a model (P3, right)

Role Switching and Handovers: Being the Modeler and Providing Information
In the studies, two patterns of work divisions could be observed: For two pairs, one participant drove the interaction and the other (mostly) provided information verbally (Fig. 5, left). In the other pairs, both partners were active and took turns in driving the interaction (Fig. 5, right). In one situation (Fig. 6), P2.2 started to interact while P2.1 observed this (left). After discussing P2.2's changes (center), P2.1 takes over and interacts with P2.2 watching (right). This handover situation is typical for cooperative work on shared resources – the surprising fact is that the handover was between normal users switching the role of a modeler.

Fig. 5. Modeler role and role change in lay modeling: One participant guiding the interaction (by pointing to an element) and the other being passive (P1, left), both participants being active (P2, center) and one participant making a global gesture to reference the whole model (right)

Role division in Modeling
Sometimes, we observed that the person interacting with the models was not the driver of interaction. For example, P1.1 explicitly told P1.2 which changes to apply without touching the displayed material. This indicates that P1.1 took the role of the modeler, while P1.2 only operated the modeling tool. Fig. 6 shows another example of this, as in addition to the description above P2.1 also gave instructions to P2.2.

Fig. 6. Handover: P2.2 changes the model (left), then P2 discuss the change (center) and P2.1 alters the model, taking over the modeler role (right). Sequence captured within 13 seconds.

These observations include behavior in which some participants took over responsibilities and tasks of expert guides during cooperation on the process models – even if they had not considered themselves to be modeling experts in the beginning of the study session (see section 3.1). This might be explained as a result of imitation, meaning that the participants used patterns of behavior they had seen from an expert in a modeling workshop before. However, our observations of role division and switching also suggest that there was more than only imitation, as these are patterns that transcend what can be observed in workshops (e.g. there would be no role switching for the expert guide and modeler), as they were not instructed to do so but patterns emerged for each pair (e.g. the way in which they divided the roles) and as they took the negotiation of model content seriously rather than doing it in playful imitation.

4.4   Limitations of Interaction with Models

Despite the promising and inspiring observations described above, we also found limitations of cooperation on process models, which suggest that for some tasks of using models, additional support such as guidance and translation is (still) required.

Securing Common Understanding
We observed some occasions in which the participants could not create a common understanding of models. Utterances like "I don't understand this at all" (P1.1) and "What do you mean by that?" (P2.1) show that explaining models was not always sufficient. In addition, we observed P4 to initially identify two elements to be different, but after a short discussion agreed on them not to be different. Thus, there is a need to structure and sustain communication content on process models as a e.g. facilitator could – this would support cooperators to sustain their understanding of processes and to note parts in which additional communication is required.

Integrating Opinions and Views into the Description
While we asked the participants to document the process as it is, some "contributed what has to happen there according to my opinion" (P3.1). This documentation of a to-be-process was not what we had asked the participants for and might not have

happened in presence of stricter guidance. This needs to be seen as a limitation of cooperation on process models without expert guidance.

Interaction Leading to Conflicts
In one situation, P1 had walked through the process, exchanged views and even reached common understanding for certain parts. However, for one specific part they started to argue. Suddenly, P1.1 told P1.2 that he completely disagreed with his view. This indicates that the discussion of gaps between views needs to be supported better, e.g. by comparing it to a third view on the process or on organizational standards.

5   Analysis: A New Perspective on Models in Cooperative Work

Looking at our observations we can differentiate ways of using models in collaboration (see Table 2): First, process models were used to structure and contextualize communication, thus becoming artifacts guiding cooperation. Second, they were used as artifacts in cooperation, supporting knowledge exchange and the development of common understanding. Third, using process models led to activities clearly related to modeling such as negotiating process content. This shows that, given a comprehensible modeling language, people can make send of model in cooperation. We are quite aware that due to the limited sample size, these results cannot be generalized. However the results may inform design on supporting lay users working on and with models.

Table 2. Process models in cooperation: Guidance, articulation and modeling

Process models guiding cooperation	Models structuring communication
	Models contextualizing communication
Process models as articulation during cooperation	Discursive interaction on process model content
	Developing a common understanding
	Transferring knowledge from different perspectives
Modeling as part of cooperative model usage	Negotiating process content
	Assuming roles that are usually related to experts

With respect to **process models guiding cooperation**, the observations described in section 4.1 show a strong relation between statements of the participants and model content. Communication aspects such as storytelling based on a model or models guiding explanations and discussions illustrate that the process models were an *integral part* of communication **on the processes**, *triggering and framing* it. We observed models triggering the start of communication on certain process steps, communication of episodes and framing communication both with respect to giving it context and focusing it to certain aspects of a process. The integration of models into communication came natural to the participants, without effort and some participants even used references to models as communication, e.g. as answers to questions. Thus, we can conclude that the process models enabled the participants to discuss process content, which shows that there is potential for the usage of such models by non-experts.

Looking at cooperative work with models as described in section 4.2, we observed models to trigger and support activities in cooperation, which transcended the discussion of certain steps, such as developing a common understanding of processes. This shows that the participants could understand the semantics of the models they worked with. Moreover, we could observe that models became the *center of cooperation* in many ways, including using them as means of knowledge transfer between different perspectives. This shows the *normative character* participants perceived in the models, as it illustrates their intention to have their perspectives integrated and expressed properly in them. It also shows that participants related a meaning to process model constructs and their sequence rather than just using it as an arbitrary graphical artifact. The fact that they used the process models as expressions in their conversations is a strong signal for the adoption of the models and, in addition to the other observations, shows how the models became central artifacts of the participants' cooperation.

The participants even engaged into modeling activities such as negotiating process content and visualizing them. This shows their capability to understand process structure, relating meaning to it and also perceiving it as an abstract visualization of real process activities. It also indicates that they are capable of relating their own experiences with the process to an abstract visualization of it. It is especially interesting that the participants also formed a modeling team as known from expert-guided modeling scenarios (see section 2.2), e.g. by taking over modeling duties, which are usually related to experts. This shows the potential of modeling by non-expert-modelers.

We also found limitations of model usage in self-directed settings. Sometimes participants were not able to secure a common understanding by just referring to the model. Some participants also tried to overpower others by integrating personal views rather than sticking to process documentation. This especially happened when there was a gap in hierarchies (cf. [9]). Situations like these require facilitation in order to allow people to express themselves freely without the fear of being overruled.

The insights stemming from our study exceed existing knowledge on models in cooperation (see section 2). They show that people can use models cooperatively even if they have not been trained for this, thus answering our first research question (**RQ1**). While this might sound trivial, to the knowledge of the authors there is no work available on the degree of cooperation with or on models that is possible for non-expert modelers. More than that, many researchers and experts even doubt that these people can work with models without expert guidance. It is also important that these results have been derived in a setting in which no model related facilitation took place. They even engaged into activities of modeling without being asked to do so thus assuming roles that are normally limited to experts (c.f. section 2.2). Therefore, we claim that lay users can use models in cooperation even without expert guidance.

In addition, it is important to understand that the findings described in this paper can clearly be attributed to the use of process models as artifacts in cooperation as opposed to arbitrary graphical artifacts. The effects we observed such as negotiation of process content or aligning communication to the sequence of the process discussed could not have been reached with arbitrary (graphical) artifacts or without an understanding of the semantics of the models used in the study.

Summing up, the following aspects can be described as minimal requirements for non-expert model interaction (**RQ1**):

- Users have to be knowledgeable about the process that is being modeled
- Co-located settings supporting communication and work with artifacts
- Reducing the semantics of the modeling language to simple constructs that allow for e.g. constructing sequences and allocating tasks

- Providing tools with a reduced feature set tailored to simple modeling tasks
- Simple means for input like direct manipulation of models (e.g. touch interface)

Considering the latter, one might be tempted to attribute some observations solely to the use of an interactive large screen and not to interaction with process models. While the environment for sure influenced cooperation, we did not describe any observation attributed mainly to the large screen. The environment rather helped us recognize participants' actions and intentions. It also showed direct manipulation through touch to be an adequate means of interaction for non-expert-modelers (**RQ2**). Thus, if the interactive large screen positively influenced people´s interaction with models, it shows that people can use models in cooperation given proper means to do so.

6   Looking Ahead: Prototypes for Lay User Modeling

In order to tap the aforementioned potentials of lay users cooperating on models and engaging into modeling activities, we developed two prototypes.

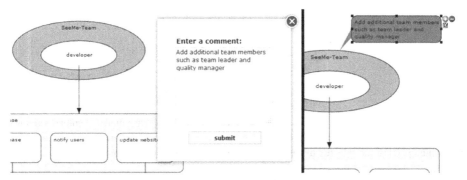

Fig. 7. A web interface allowing for commenting selected elements via a simple text-input box

As we could see, non-experts-modelers are capable of relating qualified communication to models. To tie communication about models closer to the model, thus structuring communication about them, our first prototype allows for model elements to be annotated (Fig. 7). To annotate an element the user has to select it, press the "+"-button on the top right corner of the element (see Fig. 7) and type in some text. This creates a green bubble next to the element. To stimulate the input of users, it is supported by generic prompts asking them to provide certain elements. To support follow-up communication, bubbles may again be annotated, supporting discussions on elements and model parts. Using text enables all process participants – experts and lays – to sustain communication and follow up on it in later modeling, supporting them to align communication and modeling work.

To widen the semantic available for non-experts in modeling (cf. section 5), we developed a prototype that supports what we call **meta modeling**: Selecting a single element, a user may describe what is happening in this process part by specifying suitable tags (c.f. *Search tags* on the right side of Fig. 8) or using natural language description e.g. resulting from the text input shown in Fig. 7. The tool then searches for existing models within a database, which contain the same or similar descriptions and proposes them to the user (c.f. list of *Similar models* in Fig. 8 right), including a small preview window to evaluate a model at a first glance and a full-scale preview in order to select parts out of it. If the user has found a suitable model, she may select parts of it and simply fill this selection into the previously created element by pressing the *Paste*-Button (c.f. bottom-left in Fig. 8). This way, the **meta modeling** tool allows people to document complex situations on their own while supporting them in using the complete semantics of a modeling notation without any further guidance. This supports tasks we could observe in the study such as telling stories based on a process model and detailing the model. In addition, it enables non-expert modelers to translate text input made with the prototype described above on their own.

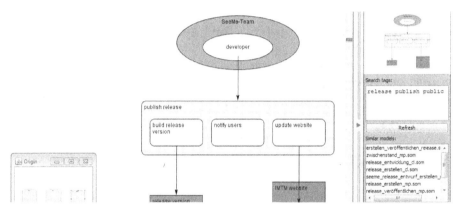

Fig. 8. The interface of the meta-modeling tool displaying a selected model (center), similar models based upon a number of tags entered by the user (right) and the original (base) model, into which parts of the selected model can be pasted (*Origin window*, bottom-left)

At the moment, these tools are prototypes but as they are well grounded in our studies, we expect them to change the way lay users can use process models in cooperation. Studies on their application will reveal if these expectations will come true.

7   Conclusion and Outlook

In this paper we show that process models can be used in cooperative work of non-expert modelers without external guidance. Our work was inspired by the lack of insights on cooperative work of lay users with process models and our experiences with using process models. In order to explore how people work with process models outside e.g. modeling workshops, we performed a study in which participants were faced with process models and tasks related to them. We could observe three basic patterns of model usage clearly directed towards and related to models; model guiding cooperation, models becoming artifacts of cooperation and cooperative modeling behavior by lay users. These observations indicate that participants adopted models as part of their cooperative work. We conclude that – given adequate means of interaction – people are capable to perform basic modeling tasks without expert guidance.

To illustrate support for this work, we presented two prototypes enabling lay user to work cooperatively with models and without expert guidance. These insights transcend existing knowledge on lay user cooperation with and on process models. It should be noted, we recognize that this work is based on one modeling language (SeeMe) so far and that it only concerns a subset of modeling tasks.

In future work we will try to find further tasks which lay users can perform with adequate support and to explore which complexity they can master and which quality of models they can produce. We will also continue analyzing qualities that model usage can bring to cooperation support. This will also be compared against other ways to support of interacting with processes such as textual descriptions or paper representations. Further work will also need to show whether the effects shown in this paper are sustainable and whether they can be leveraged systematically. As our work shows, there is vast potential in using models for cooperation support of lay users, and we are convinced it is worthwhile pursuing this work and tap this potential.

References

1. Schmidt, K., Simone, C.: Coordination mechanisms: Towards a conceptual foundation of CSCW systems design. Computer Supported Cooperative Work (CSCW) 5, 155–200 (1996)
2. Suchman, L.: Making work visible. Communications of the ACM 38 (1995)
3. Prilla, M.: Models, Social Tagging and Knowledge Management – A fruitful Combination for Process Improvement. In: Rinderle-Ma, S., Sadiq, S., Leymann, F. (eds.) BPM International Workshops 2009, Ulm, Germany (2010)
4. Prilla, M., Nolte, A.: Integrating Ordinary Users into Process Management: Towards Implementing Bottom-Up, People-Centric BPM. In: Bider, I., Halpin, T., Krogstie, J., Nurcan, S., Proper, E., Schmidt, R., Soffer, P., Wrycza, S. (eds.) EMMSAD 2012 and BPMDS 2012. LNBIP, vol. 113, pp. 182–194. Springer, Heidelberg (2012)
5. Mendling, J., Reijers, H.A., Cardoso, J.: What Makes Process Models Understandable? In: Alonso, G., Dadam, P., Rosemann, M. (eds.) BPM 2007. LNCS, vol. 4714, pp. 48–63. Springer, Heidelberg (2007)
6. Herrmann, T.: Systems Design with the Socio-Technical Walkthrough. In: Whitworth, B., de Moor, A. (eds.) Handbook of Research on Socio-Technical Design and Social Networking Systems. Information Science Reference (2009)
7. Carell, A., Nolte, A.: Seamless integration of collaborative creativity techniques into group process modelling. In: Bodker, K., Bratteteig, T., Loi, D., Robertson, T. (eds.) Proceedings of the Eleventh Conference on Participatory Design 2010, pp. 182–197. ACM, New York (2010)
8. Herrmann, T., Nolte, A., Prilla, M.: Awareness support for combining individual and collaborative process design in co-located meetings. International Journal on Computer-Supported Cooperative Work (IJCSCW). Special Issue on Awareness (2012))
9. Nolte, A., Prilla, M.: Process models as neutral ground in collaboration, but power matters. In: Nolte, A., Prilla, M. (eds.) Workshop on Collaborative Usage and Development of Models and Visualizations. CEUR-WS (2011), http://ceur-ws.org/Vol-777/
10. Herrmann, T., Hoffmann, M., Kunau, G., Loser, K.-U.: Modeling Cooperative Work: Chances and risks of structuring. In: Cooperative System Design. A Challenge for the Mobility Age (Coop 2002), pp. 53–70. IOS Press (2002)

11. Rittgen, P.: Collaborative Modeling: Roles, Activities and Team Organization. International Journal of Information System Modeling and Design (IJISMD) 1, 1–19 (2010)
12. Witschel, H.F., Hu, B., Riss, U.V., Thönssen, B., Brun, R., Martin, A., Hinkelmann, K.: A Collaborative Approach to Maturing Process-Related Knowledge. In: Hull, R., Mendling, J., Tai, S. (eds.) BPM 2010. LNCS, vol. 6336, pp. 343–358. Springer, Heidelberg (2010)
13. van der Aalst, W.M.P., ter Hofstede, A.H.M., Weske, M.: Business Process Management: A Survey. In: van der Aalst, W.M.P., ter Hofstede, A.H.M., Weske, M. (eds.) BPM 2003. LNCS, vol. 2678, pp. 1–12. Springer, Heidelberg (2003)
14. Gjersvik, R., Krogstie, J., Følstad, A.: Participatory development of enterprise process models. In: Information Modeling Methods and Methodologies, pp. 195–215 (2004)
15. Richardson, G.P., Andersen, D.F.: Teamwork in group model building. System Dynamics Review 11, 113–137 (1995)
16. Recker, J., Safrudin, N., Rosemann, M.: How Novices Model Business Processes. In: Hull, R., Mendling, J., Tai, S. (eds.) BPM 2010. LNCS, vol. 6336, pp. 29–44. Springer, Heidelberg (2010)
17. Renger, M., Kolfschoten, G.L., De Vreede, G.J.: Challenges in collaborative modelling: a literature review and research agenda. International Journal of Simulation and Process Modelling 4, 248–263 (2008)
18. Cherubini, M., Venolia, G., DeLine, R., Ko, A.J.: Let's go to the whiteboard: how and why software developers use drawings. In: Proceedings of the SIGCHI Conference on Human Factors in Computing Systems, pp. 557–566 (2007)
19. Mendling, J., Reijers, H.A., Cardoso, J.: What Makes Process Models Understandable? In: Alonso, G., Dadam, P., Rosemann, M. (eds.) BPM 2007. LNCS, vol. 4714, pp. 48–63. Springer, Heidelberg (2007)
20. Herrmann, T., Hoffmann, M.: The Metamorphoses of Workflow Projects in their Early Stages. Computer Supported Cooperative Work 14, 399–432 (2005)
21. Schmidt, K., Tellioglu, H., Wagner, I.: Asking for the moon Or model-based coordination in distributed design. In: Balka, E., Ciolfi, L., Simone, C., Tellioglu, H., Wagner, I. (eds.) ECSCW 2009: Proceedings of the 11th European Conference on Computer Supported Cooperative Work, Vienna, Austria, September 7-11 (2009)
22. Heyer, C.: High-Octane Work: The oil and gas workplace. In: Wagner, I., Tellioglu, H., Barka, E., Simone, C., Ciolfi, L. (eds.) Proceedings of ECSCW 2009. Springer, Vienna (2009)
23. Fleck, R., Fitzpatrick, G.: Teachers' and tutors' social reflection around SenseCam images. International Journal of Human-Computer Studies 67, 1024–1036 (2009)
24. Suchman, L.A.: Plans and Situated Actions: The Problem of Human-Machine Communication. Cambridge University Press (1987)
25. Pankoke-Babatz, U.: Electronic behaviour settings for CSCW. AI & Society 14, 3–30 (2000)
26. Mendling, J., Reijers, H., Recker, J.: Activity labeling in process modeling: Empirical insights and recommendations. Information Systems (2010)
27. Strauss, A.L., Corbin, J.M.: Basics of qualitative research: Techniques and procedures for developing grounded theory. Sage Publications (1998)

Designing the Software Support for Partially Virtual Communities

Francisco Gutierrez[1], Nelson Baloian[1], Sergio F. Ochoa[1], and Gustavo Zurita[2]

[1] Computer Science Department, Universidad de Chile
Av. Blanco Encalada 2120, 3rd Floor, Santiago, Chile
{frgutier,nbaloian,sochoa}@dcc.uchile.cl
[2] Control Management and Information Systems Department, Universidad de Chile
Diagonal Paraguay 257, Santiago, Chile
gzurita@fen.uchile.cl

Abstract. Designing software platforms to support the activities of partially virtual communities (PVC) is a challenging task since the supporting services must evolve continually according to the community evolution. Moreover, unsuitable supporting services usually lead the community to its demise. Therefore, these platforms must count on a flexible architecture that provides suitable services as a way to support interactions among community members, and thus contributing to keep the community sustainability. This article proposes a software architecture that helps software designers to address this challenge. Such a model can be used not only to ease the architectural design process, but also to evaluate already implemented PVC supporting systems. The article also shows a preliminary evaluation of both roles of the proposed model and discusses the obtained results.

Keywords: Social system architecture, software architecture, partially virtual communities, supporting systems.

1 Introduction

Over recent years, social computing has become present in many aspects of our daily activities. Although virtual communities have been present in several scenarios for some time, the recent rise of social computing systems has helped spread and diversify them. Several taxonomies have been proposed to classify these people associations [16, 28]. This article considers just one of these types that we have called *Partially Virtual Communities* (PVC) [11]. In such communities, members have the opportunity to interact frequently through both a virtual and a physical space. Examples of PVCs are the communities of a university course or people in a small neighborhood.

Membership in these communities is quite stable, meaning that few people join or quit these communities. PVCs depend on a certain personal interaction and knowledge among their members. Therefore, when two people decide to interact through the virtual space (i.e. the community supporting system), they know each other, and such contextual information (i.e. the mutual knowledge) allows them to appropriately interpret others' contributions. When a member makes a commitment, the rest of the participants know (or estimate) how trustworthy that commitment is,

V. Herskovic et al. (Eds.): CRIWG 2012, LNCS 7493, pp. 73–88, 2012.

based on the previous behavior of that person. The personal knowledge among members makes these communities stronger and tightly linked.

Since a PVC involves partially virtual participants, it inherits several features from physical and also from virtual communities. Although in a PVC its members cannot easily leave the community (because other links keep them connected; i.e. contractual links), several studies indicate that community members avoid participating when a certain number of conditions are not satisfied, finally leading to the community extinction [17, 25]. The unsuitability of services that support the interactions among community members usually acts as a trigger for such an end [14]. Therefore, ensuring the suitability of these services is mandatory for any platform that supports online and partially virtual communities. This service suitability is temporal, since it depends on the lifecycle stage living by the community. As long as the community evolves, some services become obsolete, requiring new ones. Software designers have to identify the services currently required by the community, and envision those eventually required in the near future, as a way to prepare the supporting platform for the next evolution stage. Thus, designers can conceive a supporting architecture that is able to evolve with the community, and avoid the software support triggers the community demise.

To the best of our knowledge, the literature does not report structural designs that help address this challenge. Therefore, designers of this type of systems must improvise or adopt ad hoc solutions to deal with this issue.

This paper proposes a software architecture that helps design the architecture of PVC supporting systems. The model can be used as a design guideline for under development solutions, and also as an instrument to measure suitability of services embedded in already implemented platforms. The proposal was used to evaluate two already implemented PVC supporting systems, and also to design a particular PVC supporting platform. The obtained results are highly encouraging.

Next section defines the concept of a PVC. Section 3 presents the related work, which is focused on requirements and design guidelines for PVC supporting systems. Section 4 discusses a list of functional and non-functional requirements that should be considered when designing these systems. Section 5 proposes the software architecture for PVC supporting systems and it shows how its components help address the requirements presented in section 4. Section 6 presents a short-term field evaluation with expert and end users. Section 7 analyzes two commercial community-supporting systems and identifies their limitations to support interaction among members of a PVC. Finally, section 8 presents the conclusions and future work.

2   Partially Virtual Communities

A partially virtual community is a hybrid between a physical and a virtual community. This classification considers just the way in which their community members interact. Therefore, we assume that members of a physical community perform just face-to-face interactions, and members of a virtual community interact only through supporting systems (e.g. email or a Web application). Clearly, most communities involve physical and virtual interactions in several percentages. The features of a hybrid community will be affected by the features of the physical and virtual communities, according to their percentages of representativeness. For example, a neighborhood community is a PVC that probably is close to a physical community, and a gamers community is a PVC that is probably close to a virtual community. In this article we consider the PVCs that are in the middle area of this spectrum (Fig. 1).

Fig. 1. Spectrum of communities according to the nature of their member interaction

There is a lack of consensus regarding an appropriate definition of the terms physical and virtual community [29]. Therefore, for physical communities we adhere to the definition given by Ramsey and Beesley, which indicates that they are a group of people who are bound together because of where they reside, work, visit or otherwise spend a continuous portion of their time [33]. Regarding online communities, we adhere to the definition of Lee et al. which indicates that they correspond to "a cyberspace supported by computer-based information technology, centered upon communication and interaction of participants to generate member-driven contents, resulting in a relationship being built up"[21]. Based on these definitions, we define a PVC as *a group of people who interact around a shared interest or goal using technology-mediated and face-to-face mechanisms.* Depending on the community context, different PVCs could involve different degrees of virtualness.

In terms of size, PVCs accomplish with the "Dunbar's Number" [7], because physical and virtual communities seem to already accomplish with it [9]. This number indicates that human social networks involve stable relationships just in a range of between 100 and 200 individuals. These relationships are stable, when an individual knows who each person is, and how each person relates to every other.

Similar to physical and virtual communities, the PVC structure is diverse and it could be complex. The complexity comes from the fact that these communities could involve social and also (formal or informal) organizational goals. Therefore, the social structure that rises spontaneously through member interaction is influenced by the organizational structure (in case that this last one is present) generating a hybrid structure that is particular for each PVC community. However, we can assume a hierarchical structure for the PVC due it is basis of a social group [3]. In fact, whenever a group of people interacts within a community, a leader-follower relationship almost always emerges [38]. Therefore, we preliminary assume a leader-follower structure for a PVC where it is possible to identify several roles, such as consumers, contributors, lurkers and veterans [36].

3   Related Work

This section presents the main requirements to be considered by designers of software systems that support activities of a PVC. We then present and discuss the existing guidelines to model these supporting platforms.

3.1   Requirements for PVC Supporting Systems

PVC platforms typically support information dissemination, self-service transactions, communication and mediation [8]. The large amount of software to support online communities that exists today, may lead to misunderstand that the development of PVC platforms for particular purposes is straightforward [4].

McMillan and Chavis [22] state there are four elements that define a sense of community: *membership, influence, integration and fulfillment of needs*, and a *shared emotional connection*. Therefore, encouraging participation and allowing social interactions lay within the basic requirements to be fulfilled by PVC platforms [39], whose design has to be driven by usability and sociability [31].

A persistent and updated identity triggers cooperation, as community members tend to identify each other and keep a track of their behavior in the past [19]. Moreover, user behavior and information published under personal profiles allows community members to infer relationships and/or other data related to different users [24]. Lee et al. have identified a set of requirements that can be used to foster social interaction: *common ground, awareness, social interaction mechanisms* and *place-making* [20]. Information sharing, knowledge of group activity, and coordination are central to successful collaboration [6]. Collaborative systems like PVC platforms should consider context to support interaction among group members. In fact, users especially value information related to status and physical location, as well as profile information [13].

Although all these functional requirements (FR) identified in the literature are relevant in the design process, establishing the non-functional requirements (NFR) is also highly relevant to obtain a design that helps keep the community alive. For example, scalability of these platforms is important since they usually provide support to several communities.

It is well known that the most effective way to address the NFR in a software system is considering them in its architectural design [35, 40]. Such an architecture must integrate harmoniously all FRs and NFRs of the system which, per se, is a challenge due to the interrelationships existing among these requirements. Moreover, the services provided by the architecture must be suitable for the end-users, particularly in PVCs where the members' interactions are based on a voluntary use of the supporting system.

3.2   Guidelines to Design PVC Supporting Systems

To the best of our knowledge, there are no particular proposals to help design the architecture of PVC supporting systems. However, there are some results from online communities studies, which should be considered when modeling these systems. For example, Preece and Shneiderman [32] have identified that community members are relatively shy at first, typically evolving from *readers* (passive stage) to *leaders* (active stage). Therefore, supporting services provided by a PVC platform must consider this user behavior evolution.

Similarly, Kim [18] studied users in online communities and defined some guidelines, such as defining a community purpose, developing spaces for interaction, and creating meaningful profiles that may evolve in time. Porter [30] presents the *AOF Method* (activities, objects, features), which consists on a prioritization scheme for designing social Web applications, and a model of five stages of the usage lifecycle.

Gutierrez et al. state that participation is a key metric to evaluate the success of an online community [10]. Based on that premise, they propose a framework for enabling interaction among users. The framework models virtual communities in three sections: (1) services that allow interaction, (2) participation and motivation strategies, and (3) definition of the software platform through which the community is going to interact.

Howard proposes a model to address the community member behavior and tries to identify the services required by them [15]. This model is based on four components: *remuneration, influence, belonging* and *significance*.

Concerning guidelines for social platforms, Crumlish [5] identifies a series of social interface design patterns and analyzes how they are applied into different systems. Van Duyne et al. [37] present a pattern for designing online communities, considering policies, moderation, anonymity, interaction, trust, sociability, growth and sustainability. These patterns provide a partial solution to the design of PVCs, because they lack of support for physical interactions required by PVC members.

The literature also reports an ample variety of architectural and design patterns that were not particularly proposed to model PVCs, but that could be used as general guidelines for it. For example, Schümmer and Lukosch define a pattern language for computer-mediated interaction [34] that can be used to design several aspects of the community support, such as users identification, contacts (buddy list) and mechanisms for reciprocity and rewards among community members.

4   Requirements to Support PVC Activities

This section identifies FR and NFR that are usually present in this type of supporting platforms. These requirements have been obtained from the literature review and from the authors past experiences as designers of these software platforms.

Typically, PVC platforms are Web applications either open to public members or closed in private groups or organizations. The context that defines the community will state how information will flow outside its borders. For example, when the system must support inter-organization processes, interoperability should be considered as a mandatory requirement [1].

These systems should implement at least two roles: *admin* and *standard* users. The admin-user takes the role of community manager, with permission to coordinate and control participation and membership. This role contributes to keep the community governance within a certain suitability range and may be a way of responding to the perceived lack of strong governance structures in online communities [27].

When designing the interaction space, the supporting system should consider two disjointed environments: *public* and *private* [26]. Sharing resources between these two environments has to be possible. Public spaces foster communication throughout the community, and private spaces allow users to organize their personal information, as well as interact and share content with others.

The platform architecture should also consider services that allow synchronous and asynchronous communication among community members [26]. It has to support three different kinds of interaction: user-to-user, user-to-a selected group and user-to-community. Counting on these strategies provides flexibility to user participation. Awareness about the members' availability usually helps to promote these interactions. Since the community is partially physical, user location awareness mechanisms should be considered to trigger face-to-face interactions.

Concerning the NFR for PVC supporting systems, the most relevant and common ones seem to be: *performance, uptime, maintainability* and *scalability*. These requirements try to address the services usability (particularly the first two NFR) and the platform evolution. Other requirements such as *privacy* and *security* have also to be taken into consideration. Finally, in order to ensure member satisfaction towards the system, as well as effectiveness and efficiency when supporting user interaction, the software support has to comply general *usability* principles. Table 1 summarizes the requirements to model of a PVC supporting system.

Table 1. Requirements for a PVC supporting system

Req.	Description
FR 01	The system should provide registration mechanisms that facilitate the appropriation of the platform by users [5, 15, 18, 30].
FR 02	The system should provide mechanisms for managing a personal identity by users [5, 8, 18, 30].
FR 03	The system should include awareness mechanisms in the form of users' availability, action identification and notifications [5, 6, 13].
FR 04	The system may include location awareness in order to allow face-to-face interactions and break the barriers linked to virtualness [5, 13].
FR 05	The system should allow and trigger relationship building among community members; e.g. friends, circles, groups [5, 8].
FR 06	The system may provide services for sharing content and media with other users, either in private groups or publicly [5, 10].
FR 07	The system should provide interaction mechanisms, such as synchronous and asynchronous communication modules [5, 18, 26, 37].
FR 08	The system may provide mechanisms for supporting collaboration and content creation among community members [5, 10].
FR 09	The system should include control mechanisms, such as peer moderation, governance structures and filters [5, 27, 37].
FR 10	The system should be designed following a motivation and participation strategy in order to ensure a certain level of activity through time [4, 8, 10, 32].
NFR 01	The system should react to short response times against any request made by users or its components [23].
NFR 02	The system should be highly available (uptime), since PVCs are supposed to break down time barriers, allowing members to interact at any time [23].
NFR 03	The system should be maintainable and extensible, because communities evolve naturally in time and follow a specific lifecycle, as well as its users [2].
NFR 04	The system should be scalable, since it has to be able to handle a continuous growing number of users and contributions made within the community [14].
NFR 05	The system should ensure privacy and security, as PVCs have to be trustworthy for users in order to trigger interactions [5, 11].
NFR 06	The system should be usable, since it has to support community member interaction and deal with different kinds of users [11, 31].

5   Software Architecture for PVC Supporting Systems

Herskovic et al. state that requirements of collaboration systems should be layered [12]. Requirements in the upper layers are highly visible to users and developers, because they represent services that are exposed to end-users through the application front-end. Following this line of reasoning, we propose a software architecture composed of three layers (Fig. 2): *user, interaction* and *community* layer. The *User Layer* refers to specific actions to be performed by a single user within the community. Some of the expected tasks to be carried out by a user are logging into the software and managing his/her profile and personal identity. The *Interaction Layer* refers to all actions and services to be done by two or more users, or with the intention of causing an effect on the community. The *Community Layer* refers to the global scope of the community, the elements that define the software, and all the principles that directly affect the whole group.

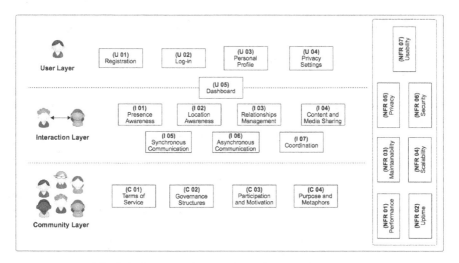

Fig. 2. Software architecture for PVC supporting systems

The User Layer is composed of five services; one of them is shared with the Interaction Layer. The *registration, log-in, personal profile,* and *privacy settings* manage the identity and visibility of a single community member. The *dashboard* is where personal contributions are published alongside those of the other members. It allows filling-up the feedback loop of information where personal and public notifications foster interaction among users.

The Interaction Layer is composed of seven services: *presence awareness, location awareness, relationships management, content and media sharing, synchronous* and *asynchronous communication,* and *coordination.* The two requirements related to awareness are justified because of the need of users to foster face-to-face interactions, as well as requirements linked to services providing different communication channels for users interaction, e.g. a message board or a chat room. The relationships management component is a key issue in this architecture. Such a service allows users to identify other members and send an interaction request to them. The coordination service regulates the access to shared resources of the community (e.g. shared object or the communication channel). The content and media sharing component is closely linked to participation in communities that are based in collaborative work. Using such a service, users may interact with each other to contribute or create new content, thus leading the community to evolve.

In the Community Layer we can identify the four mechanisms (rather than proper software services) that define the context where a community lives and evolves in time. These mechanisms are: *terms of service, governance structures, participation and motivation strategies,* and the *purpose* and linked *metaphors* to be used when designing the community. In particular, this layer is usually invisible to end-users, because its components affect the whole structure of a community. However, it is the one that has the greatest impact in the design of PVC supporting systems.

The complexity of architecture presented in Fig. 2 and the nature of these supporting applications indicate that these systems must be framed in a client-server architecture, where the user layer lives in the client side, and the two lower layers are in the server side. This design decision simplifies the services implementation.

Concerning the NFRs, they are "transversal requirements"; therefore, they affect all the services provided through the architecture. The proposed architecture considers these NFR and proposes mechanisms to address them. Particularly, the identification of services and their separation by concerns (i.e. user, interaction and community) make the systems *maintainable* and *extensible*. This property comes from structuring the systems using layers [2]. We can also expect an appropriate *performance* of the systems that are implemented using this architecture because it is client-server and involves just three layers [23]. Since the two lower layers (which are affected by the number of communities and users to be supported) live in the server, we can ensure the system *scalability* by increasing the computing power in the server side. The system *uptime* cannot be ensured through this architecture since it does not consider replicated components in the server side [23]; however it should be interesting to include it in the future. However, the proposed architecture partially addresses such a NFR through the use of asynchronous interaction services.

User *privacy* preferences are stored by the system; therefore the services provided by the platform must self-configure to adhere to the user privacy settings. Since this information is kept in a dual-synchronized way (i.e. in the client and also in the server), it cannot be modified unless the user has a simultaneous access to both copies of such information. This information management policy is used also to manage the personal and login information. This mechanism contributes to build *secure* systems. In addition, the architecture considers users authentication. Similar to any other domain specific software architectures, this proposal addresses the systems *usability* just accomplishing with all previous requirements (including FR and NFR).

6   Using the Proposed Architecture

In order to determine if the services considered in the software architecture are suitable to support a real PVC, we have developed the supporting system for an existing community. This community was composed of 30 students of an introductory Information Technology course of the Business School at the University of Chile. Students taking part of this experience were volunteers and were required to register and validate their accounts. They were also asked to fill up their personal spaces and publish, rate and comment discussion topics related to the course contents.

The lecturer and two teaching assistants also became community members and discussed with the other students. The users participate through an avatar to keep anonymous their interactions. The community had a manager (an external person) who tracked the interactions and gave regularly feedback to members about their behavior in the platform. This community is still in service, but the tracking period was limited to 8 weeks from its initial launch. After that period we applied a survey to end-users to gather their feeling about the usefulness of the services provided by the platform. These services were completely aligned with the software architecture. After such a validation process we carried out a focus group with six software designers: two with experience in the design of social platforms, two with experience evaluating usability of software interfaces, and the last two with no prior knowledge about modeling PVC supporting platforms. The focus group served to discuss and clarify the designers' opinion about the suitability of services and pertinence of the NFR considered in the proposal. Each designer filled up the survey and a section asking for the suitability of the NFR considered. Using these results we tried to answer the following questions:

(Q1) Are the services considered in the architecture useful to support the interactions among PVC members?
(Q2) Is the architecture a guide to design PVC supporting systems?
(Q3) Is the architecture useful to evaluate already implemented PVC supporting platforms?

Next we briefly describe the survey. Then we present the results obtained in the experimentation process with the users of the PVC platform (section 6.2), and also those gathered in the focus group with software designers (section 6.3).

6.1   Survey

The survey included an item for each service of the proposed architecture. Users rated the usefulness of such services using a 5-point Likert scale. Values of 1 and 2

Table 2. Description of supporting services

Service	Description
Personal Profile	Users have a personal space where they can manage their virtual identity. It provides support for an avatar, personal status or interests.
Privacy Settings	Users can decide what information will remain public and private. Also, they manage how they will receive notifications (e.g. email, in-site).
Dashboard	A main page where is published automatically the recent activity in the community, such as new messages and recent contributions.
Presence Awareness	Users can see the list of the other community members that are currently logged-in into the platform.
Location Awareness	Users can indicate their location by choosing a place from a list of options. If there are two users at the same place and time, they will receive a notification according to their privacy settings.
Relationships Management	Users can specify relationships among them, such as being part of a same group or being friends. This requires symmetric validation.
Content and Media Sharing	The system supports media uploading (e.g. documents, pictures and videos), classifies it into categories and allows users to comment on them.
Synchronous Communication	The platform supports a video chat room for logged-in users. They have to allow camera and microphone access beforehand.
Asynchronous Communication	Users can publish, comment and rate discussions related to the different topics they have worked on the lecture sessions.
Coordination	The system provides a calendar with different permission levels: users can schedule activities that are private, or involve groups.

correspond to "negligible" services, a value of 3 corresponds to a "desirable" service, and a value of 4 or 5 means the service is "mandatory". The survey also included an open comments section where users could suggest services to improve the system.

Some services considered in the model, such as registration and identification, were not considered in the survey, since they are either used only once, or required to access to the software support. Similarly, terms of service, governance structures, motivation and participation, and purpose and metaphors were also left out because they are invisible to end-users. Table 2 summarizes the services considered in the survey.

6.2   Users Perception versus Designers Perception

Fig. 3 shows the usefulness of each service according to users and designers. Dark bars represent the average value assigned by the users to the usefulness of such services. Light ones show the numerical representation of the *usefulness perception level* according to the designers' opinion. A continue scale from 0 to 10 was used to represent the usefulness of each service.

According to results shown in Fig. 3, most services were useful for the community members. Moreover, the usefulness assigned by the end-users was similar to the ones assigned by the software designers. Analyzing the results and also the students' comments in the survey, we have identified some problems in the services implementation. Services like synchronous communication and coordination were not suitably implemented in the PVC supporting system. Therefore there is an important gap between the expected and the perceived value of such services.

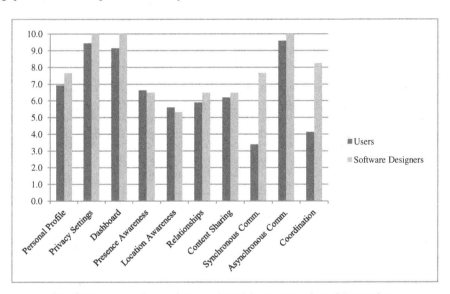

Fig. 3. Users vs. software designers' usefulness perception of the services

The spontaneous responses given by five end-users into the survey indicate that they would have preferred a simple chat room instead of the video-chat embedded in the system. This reflects that the community in fact requires this service, but it was not implemented properly. Concerning the coordination service, the panel of software designers agrees that a service that implements coordination mechanisms is desirable for this kind of community. However, end-users assigned a usefulness value considerably lower than expected one. This was also reflected on spontaneous comments that end-users stated at the end of the survey. The comments show a lack of initiative to use such service since it was not mandatory to perform the community activities during the experimentation period. The use of this service by either the community manager or other users would have motivated that community members use it consequently. Figure 4 shows the declared and the perceived usefulness assigned by end-users. Dark bars are the result of using the current implementation of a service. Light bars represent the value of each service (according to users' opinion) when they are properly implemented.

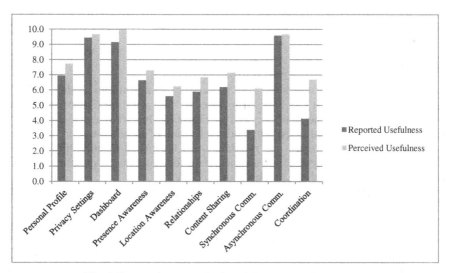

Fig. 4. Reported vs. perceived usefulness according to users

These results show that all services considered in the software architecture are considered useful by the end-users. This preliminary conclusion provides a first response to the Q1. In case of identifying a gap between the reported and the expected usefulness of a service, the cause can be: (1) inappropriate service implementation (that is the case of the synchronous communication mechanism), and (2) lack of initiative for using the service (that is the case of the coordination service).

6.3   Focus Group with Software Designers

According to the designers' opinion, the FR and NFR were appropriately considered in the design of the PVC supporting system. These engineers also highlighted the simplicity of the software architecture, which make it usable for many people. They were able to quickly understand the separation of concerns represented by the three layers architecture. Five participants pointed out that this model is almost complete, as it lacks just of support for activity awareness. This provides a preliminary response to Q2, which is also supported by the results shown in Fig. 3. All designers considered this architecture to be useful for analyzing other services in different contexts, and not difficult to learn.

Moreover, they think that the architecture could be used to evaluate already implemented PVC supporting systems. This provides a preliminary response to Q3.

7   Analysis of Already Implemented PVC Supporting Systems

In order to show how the proposed architecture can be applied in practice, we will briefly analyze two commercial PVC supporting systems: *Facebook* and *U-Cursos*. In this analysis we will try attempt to verify whether or not these systems satisfy the set of requirements specified in section 3, and also if the non-addressed requirements are required by the community members. Thus, we intend to show that this proposal can be used for: (1) designing new systems, (2) choose an already implemented system from a set of possibilities, and (3) identify further customizations or extensions needed to include in a supporting system that is currently used by a specific community.

7.1   Facebook

Facebook is considered as one of the most successful social platforms. Although this is a general social system, it can be used to support PVC with the *Groups* feature. A *Facebook Group* offers the same services as Facebook, but restricted to a particular group of users. Membership, visibility and moderation of these groups are supported by one or more *group admins*, and *standard users* are linked together through their own Facebook profiles. Fig. 5 shows a typical page for a user and it identifies the main components that match with the proposed architecture.

We can see most services considered in the architecture are part of Facebook. However, two services usually required by PVC were not included: location awareness (i.e. positioning of community members) and coordination mechanisms (e.g. community agenda or community members commitments). Only location awareness is partially supported by the use of geo-tagging, and there are no simple mechanisms for coordinating community members and activities. This result is not surprising because Facebook was not particularly designed to support PVCs. However, this fact allows us to show that the proposed architecture can be used as a reference to identify mandatory services in PVC supporting systems.

7.2   U-Cursos

U-Cursos is a PVC supporting system developed at the University of Chile for managing courses and fostering interaction among courses participants: lecturers, teaching assistants and students. Currently, this platform is commercial.

In the system, each course defines a specific context in the form of an independent community. Interaction is achieved through asynchronous communication (email and a discussion forum), and community members may upload and download class material and related media content. Fig. 6 shows the main user interface of *U-Cursos*.

The *U-Cursos* limitations come from the system conception. This tool was not initially designed to support PVCs, but it was evolving over time up to a tool that plays such role. Therefore, the required support for the community members' activities is still incomplete. For example, the system lacks of services that stimulate interaction

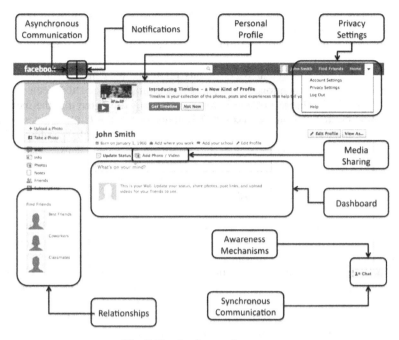

Fig. 5. Facebook page for a user

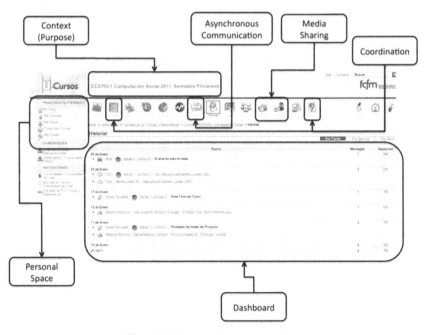

Fig. 6. U-Cursos main user interface

between users. Moreover, there is not a proper participation strategy that would eventually transform this information system into a proper PVC supporting system. The platform includes several coordination services, but it still does not support location or presence awareness.

Since the authors are regular users of this platform, we can confirm the need to count on the previously mentioned services. These limitations have also been discussed with the engineers in charge of this platform evolution, who agree that the mentioned services must be included in the system. Hypothetically, if the *U-Cursos* design were based on the proposed architecture, the implemented and also the pending services would be indentified in an early stage of the system development.

8   Conclusions and Future Work

A PVC is a hybrid between a physical and a virtual community, and we define a PVC *a group of people who interact around a shared interest or goal using technology-mediated and face-to-face mechanisms.*

This article identifies a list of recurrent requirements that should be considered when designing the architecture of a PVC supporting system. These requirements come from a literature review and also from the authors' experience developing and evaluating PVC supporting systems. Based on those requirements, a preliminary architectural model was proposed as a reference for under development and already implemented PVC supporting systems. The architecture is layered which contributes to the software maintainability and extensibility. This capability is important in these systems because they are in constant evolution. A number of recurrent services were identified as part of the architecture layers.

In order to perform a preliminary validation of the proposal, we carried out a short-term study with participants in an undergraduate course at the University of Chile. We have also conducted a focus group with expert users to examine the real and the expected usefulness of the services considered in the architecture. Such activity allows us to envision that the proposal could be used for evaluating the design of already implemented PVC software platforms.

Three research questions were stated in this article. Answering these questions will require evaluating the proposal more in-depth. However, the interim results provide a first response, which indicates the proposed software architecture considers services useful to support interaction among members of a PVC (*Q1*). This would also be useful to support the design of these systems (*Q2*) and the evaluation already implemented platforms (*Q3*).

Trying to answer the research questions 2 and 3 we have analyzed two commercial PVC supporting systems. The analysis was done through a focus group with software designers. The preliminary results indicate the proposed architecture can be used to (1) help software designers to model new PVC supporting systems, (2) identify suitable alternatives from a set of already implemented systems, and (3) determine mandatory services to be included in systems that are into production. The next steps in this work consider to conduct a survey to members of various PVCs in order to determine if the services identified in the software architecture are complete and also if all of them are mandatory.

Acknowledgements. This work has been partially supported by the Fondecyt Project (Chile), grant: 1120207, and also the LACCIR project, grant: R1210LAC002.

References

1. Benghazi, K., Noguera, M., Rodríguez-Domínguez, C., Pelegrina, A.B., Garrido, J.L.: Real-time Web Services Orchestration and Choreography. In: Proc. of the CAiSE 2010 Workshop EOMAS 2010, Hammamet, Tunisia, pp. 142–153 (2010)
2. Buschmann, F., Meunier, R., Rohnert, H., Sommerlad, P., Stal, M.: Pattern-Oriented Software Architecture: A System of Patterns. Wiley (1996)
3. Chase, I.D.: Social Process and Hierarchy Formation in Small Groups: A Comparative Perspective. American Sociological Review 45(6), 905–924 (1980)
4. Cheng, R., Vassileva, J.: Adaptive Reward Mechanism for Sustainable Online Learning Community. In: Proc. of the 2005 Conference on Artificial Intelligence in Education (2005)
5. Crumlish, C., Malone, E.: Designing Social Interfaces. O'Reilly, Sebastopol (2009)
6. Dourish, P., Bellotti, V.: Awareness and Coordination in Shared Workspaces. In: Proc. of the 1992 ACM Conference on Computer-Supported Cooperative Work (1992)
7. Dunbar, R.I.M.: Neocortex Size as a Constraint on Group Size in Primates. Journal of Human Evolution 22(6), 469–493 (1992)
8. Girgensohn, A., Lee, A.: Making Web Sites Be Places for Social Interaction. In: Proc. of the 2002 ACM Conference on CSCW. ACM Press, New Orleans (2002)
9. Gonçalves, B., Perra, N., Vespignani, A.: Modeling Users' Activity on Twitter Networks: Validation of Dunbar's Number. PLoS One 6(8) (2011)
10. Gutierrez, F., Baloian, N., Zurita, G.: Boosting Participation in Virtual Communities. In: Vivacqua, A.S., Gutwin, C., Borges, M.R.S. (eds.) CRIWG 2011. LNCS, vol. 6969, pp. 14–29. Springer, Heidelberg (2011)
11. Gutierrez, F., Baloian, N., Ochoa, S.F., Zurita, G.: A Conceptual Model to Design Partially Virtual Communities. In: Proc. of the 16th IEEE Int. Conf. on Computer Supported Cooperative Work in Design (CSCWD 2012), Wuhan, China, May 23 -25 (2012)
12. Herskovic, V., Ochoa, S.F., Pino, J.A., Neyem, A.: The Iceberg Effect: Behind the User Interface of Mobile Collaborative Systems. Journal of Universal Computer Science 17(2), 183–202 (2011)
13. Herskovic, V., Neyem, A., Ochoa, S.F., Pino, J.A., Antunes, P.: Understanding Presence Awareness Information Needs Among Engineering Students. In: Proc. of the 16th IEEE Int. Conf. on Comp. Sup. Cooperative Work in Design (CSCWD 2012), China, May 23-25 (2012)
14. Hill, T., Supakkul, S., Chung, L.: Confirming and Reconfirming Architectural Decisions on Scalability: A Goal-Driven Simulation Approach. In: Meersman, R., Herrero, P., Dillon, T. (eds.) OTM 2009 Workshops. LNCS, vol. 5872, pp. 327–336. Springer, Heidelberg (2009)
15. Howard, T.: Design to Thrive: Creating Social Networks and Online Communities that Last. Morgan Kaufmann, San Francisco (2010)
16. Hunter, M.G., Stockdale, R.: Taxonomy of online communities: Ownership and value propositions. In: Proc. of the 42nd IEEE Hawaii Int. Conf. on System Sciences (2009)
17. Iriberri, A., Leroy, G.: A life-cycle perspective on online community success. ACM Computing Surveys 41(2) (2009)
18. Kim, A.J.: Community Building on the Web. Peachpit Press, Berkeley (2000)

19. Kollock, P.: Design Principles for Online Communities. In: Proc. of the Harvard Conference on the Internet and Society (1996)
20. Lee, A., Danis, C., Miller, T., Jung, Y.: Fostering Social Interaction in Online Spaces. In: Proc. of INTERACT 2001 (2001)
21. Lee, F.S., Vogel, D., Moez, L.: Virtual Community Informatics: A Review and Research Agenda. Journal of Information Technology Theory and Application 5(1), 47–61 (2003)
22. McMillan, D.W., Chavis, D.M.: Sense of Community: A Definition and Theory. Journal of Community Psychology 14(1), 6–23 (1986)
23. Menascé, D.A., Almeida, V.A.F.: Capacity Planning for Web Services: Metrics, Models and Methods. Prentice Hall, Upper Saddle River (2001)
24. Mislove, A., Viswanath, B., Gummadi, K.P., Druschel, P.: You Are Who You Know: Inferring User Profiles in Online Social Networks. In: Proc. of the WSDM 2010, USA (2010)
25. Mousavidin, E., Goel, L.: A life cycle model of virtual communities. In: Proc. of the 42nd IEEE Hawaii International Conference on System Sciences, HICSS 2009 (2009)
26. Neyem, A., Ochoa, S.F., Pino, J.A.: A Patterns System to Coordinate Mobile Groupware Applications. Group Decision and Negotiation 20(5), 563–592 (2011)
27. Parameswaran, M.: Social Computing: An Overview. Communications of the Association for Information Systems (AIS) 19, 762–780 (2007)
28. Plant, R.: Online communities. Technology in Society 26, 51–65 (2004)
29. Porter, C.E.: A Typology of Virtual Communities: A Multi-Disciplinary Foundation for Future Research. Journal of Computer-Mediated Communication 10(1) (2004)
30. Porter, J.: Designing for the Social Web. New Riders, Berkeley (2008)
31. Preece, J.: Sociability and Usability in Online Communities: Determining and Measuring Success. Behavior and Information Technology Journal 20(5), 347–356 (2001)
32. Preece, J., Shneiderman, B.: The reader-to-leader framework: motivating technology-mediated social participation. AIS Trans. on Human-Computer Interaction 1(1) (2009)
33. Ramsey, D., Beesley, K.B.: 'Perimeteritis' and rural health in Manitoba, Canada: Perspectives from rural healthcare managers. Rural and Remote Health 7, 850 (2007)
34. Schümmer, T., Lukosch, S.: Patterns for Computed-Mediated Interaction. John Wiley & Sons, Chichester (2007)
35. Supakkul, S., Hill, T., Chung, L., Tun, T.T., Leite, J.C.S.P.: An NFR Pattern Approach to Dealing with NFRs. In: Proc. of Requirements Engineering (RE 2010). IEEE Press (2010)
36. Tedjamulia, S., Olsen, D., Dean, D., Albrecht, C.: Motivating Content Contributions to Online Communities: Toward a More Comprehensive Theory. In: Proc. of IEEE Hawaii International Conference on System Sciences (2005)
37. Van Duyne, D.K., Landay, J.A., Hong, K.I.: The Design of Sites: Patterns for Creating Winning Web Sites. Prentice Hall, Upper Saddle River (2006)
38. Van Vugt, M., De Cremer, D.: Leadership in Social Dilemmas: Social Identification Effects on Collective Actions in Public Goods. Journal of Personality and Social Psychology 76(4), 587–599 (1999)
39. Westerlund, M., Rajala, R., Nykänen, K., Järvensivu, T.: Trust and commitment in social networking - Lessons learned from two empirical studies. In: Proc. of the 25th IMP Conference, Marseille, France (2009)
40. Xu, L., Ziv, H., Richardson, D., Liu, Z.: Towards modeling non-functional requirements in software architecture. In: Proc. of Aspect-Oriented Software Design (AOSD 2005); Workshop on Aspect-Oriented Requirements Engineering and Architecture Design (2005)

Supporting Social Tasks of Individuals:
A Matter of Access to Cooperation Systems

Michael Prilla

Information and Technology Management, IAW, University of Bochum
Universitätsstr. 150, 44780 Bochum, Germany
michael.prilla@rub.de

Abstract. Today, people use cooperation systems with many different devices and interfaces. Popular systems such as Twitter illustrate this, as they can be used with many devices, provide numerous interfaces and can be integrated into many systems and web pages. As smaller cooperation systems might also benefit from such opportunities, this paper introduces the concept of 'access' to capture the different ways to interact with systems and argues that access should be regarded as a major factor for the design of cooperation systems. It understands access as vehicle to support users in carrying out their social tasks in a way that fits their needs, thus choosing from a variety of means to access systems. From an analysis of related work and of four cases of access design, it describes initial insights into influencing factors and design qualities of access.

1 Access to Cooperation Systems

Cooperation systems become more and more integrated into daily life: Systems for private communication such as Twitter are used during work and professional communication such as office e-mail is done during private time as we take them home on mobile devices. This is due to a multitude of ways to **access** such systems, including **devices** (mobile phones, desktop PCs etc.), **interfaces** (standalone, integrated in websites etc.) and **feature sets** (full functionality, simple input etc.). Thus, cooperation systems pervade daily life more than ever and "digital technology is no longer confined to a support role; it is integral to many activities" [1].

The communication service Twitter exemplifies this: It can be accessed with multiple devices and there are different interfaces and apps for Twitter, including websites, phone apps and desktop applications like TweetDeck[1]. These interfaces provide different sets of features: For example, Twitter apps on mobile phones range from simple text input to a complete set of features for using the system. This allows users to embed Twitter into daily tasks, being aware what others do and even communicating while they are riding the bus. This makes systems like Twitter attractive, usable and useful for many contexts. This example also shows that accessing systems in multiple

---

[1] TweetDeck is available at http://www.tweetdeck.com and enables access to tweets from multiple Twitter users at the same time, providing awareness on what others do.

V. Herskovic et al. (Eds.): CRIWG 2012, LNCS 7493, pp. 89–96, 2012.

ways is relevant for an individual to perform **social tasks** (communication, cooperation; see also [2]). For using other (smaller) cooperation systems, this might also be helpful, but most systems do not offer various ways to use them. This paper uses the concept of **access** to subsume devices, interfaces and feature sets to use cooperation systems and investigates its role for these systems, understanding it as **a vehicle supporting social tasks of individuals** conducted in cooperation systems. It aims to shed light on two research questions related to this lack: (1) What influences the choice of access to cooperation systems and (2) how can this influence be supported systematically.

The concept of access is discussed mainly in the context of enabling people with special needs to overcome limitations with approaches such as 'accessible design' [3]. This paper widens this understanding, using access to describe means to use the same cooperation system in different contexts. It builds on a literature analysis of related work (section 2) and an exploratory analysis of four cases the author was involved in, which also included the design of access (section 3). From this, factors influencing the adequacy of access (research question 1, section 4) and design qualities of access (research question 1, section 5) are derived.

2   Related Work

There are many theories on the interrelation between behavior and technology usage, including Activity Theory [4], Structuration Theory [5], "technology-in-practice" [6] and "embodiment" [7]. They explain how technology usage shapes human actions [4, 5], how the shape of technology depends on human action [5, 6] and how technology usage depends on the context human action is embedded in [7, 8]. They understand technology as a part of human interaction – access can be seen as an enabler for this interaction. However, they do not differentiate between layers of technology usage. Access, in contrast, includes two layers: (1) technologies used to access cooperation systems and (2) technology forming the cooperation system itself.

In interaction design, an interesting approach can be found in "use qualities", which are "properties in digital design that are experienced in use and the designer can influence at design time" [9]. This focuses technology choice on qualities needed in human interaction. These qualities include the purpose of use (with whom or for which task technology is used), the context of use (where and when technology is used) and subjective meaning (emotions and associations related to devices) [10]. While this supports access design, it does not include the choice of access to the same cooperation system depending on the context its users are in.

Approaches like media choice theories (e.g. [11]) and technology acceptance [12] do not support access design – media choice does not differentiate means to use the same medium, and technology acceptance regards utility and usability, but does not differentiate between a base technology and access to it. A promising approach are "interaction acts" [13], which decouple interactions from their implementation, supporting the creation of the same interaction with different interfaces. An approach systematically covering access to systems as discussed here, however, is still missing.

3   Access to Small Scale Cooperation Systems: Four Cases

To explore the role of access for smaller cooperation systems and to derive initial insights to build further work on, below four cases including the design of access to cooperation systems are analyzed.

3.1   Case 1: Cooperation and Awareness in a Knowledge Management System

In work with the web based learning system Kolumbus 2 (we will refer to this case as **case 1** below), we found that students perceived logging into the system and looking for new content or discussing with others to take too much time, given only a few changes per day [14]. To reduce their effort, we developed widgets offering content and system features in websites such as Netvibes and iGoogle (Fig. 1 left; see [14]). This way, users could be aware of activity in the system in parallel to other tasks. The widgets provide focused access to Kolumbus 2, including carefully chosen features. The students liked this way of using the system as part of other activities.

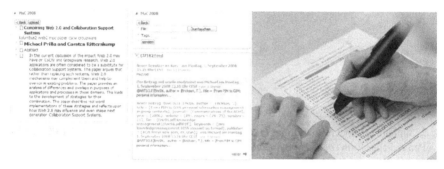

Fig. 1. Widgets providing content and features of the Kolumbus 2 platform in other websites (left) and Pen&Paper technology for accessing a service order system (right)

3.2   Case 2: Service Coordination for Elderly People

In a recent project, the goal was to support elderly people to order services supporting them in daily tasks from home[2] (**case 2**, see [15] for details). To enable access to the ordering system supporting the cooperation of the seniors and staff of an agency coordinating the services (e.g. sending clients on a shopping trip together), we used Pen&Paper technology, which uses a pen with a digital camera and paper with an almost invisible pattern to capture handwriting digitally (see Fig. 1, right). Using this technology, we created five paper forms to order 20 services. After clients filled out the forms, they were sent to the agency server and appeared in the order system.

Pen&Paper was used by the elderly clients rather intuitively: Clients were visited once for a short introduction to the pen and paper equipment. After that, they reported

---

[2] This work is part of the project „service4home" funded by the German Federal Ministry of Education and Research. Details can be found at http://service4home.net (German website).

hardly any problems with using the technology in a period of six months. Agency staff, in turn, was pleased to get structured data for service coordination and that they could use more time to engage with the client during service conduction. This case shows how a cooperative processes can be enhanced by providing a user group with proper access to the system supporting the part of the process they are involved in.

3.3   Case 3: Support for Collaborative Reflection at the Workplace

In current work on reflection in healthcare (**case 3**, [16]), we provide medical staff with data on past work and help them to re-assess this work in order to learn from it. In one scenario, we support physicians in reflecting talks with relatives (there is no special training for this often very stressful and emotional task) by providing a tool to exchange documented talks and reflect on them together.

Fig. 2. Support for documentation and reflection of talks at a hospital on a tablet PC (left) or via a Pen & Paper form (right), both including casual emotion assessment

This tool needs to support different settings, as reflection in healthcare also happens in meetings or between tasks, e.g. in shift handovers or breaks. To support this, we implemented different variants of the tool: Some physicians preferred a solution for tablet devices (Fig. 2, left) and others did not like typing on these devices and were provided with a Pen&Paper solution for the documentation of talks (Fig. 2, center). The physicians embraced the mobility of both solutions, which suited the needs of their work well. We also wanted to capture emotions related to talks, as physicians need to understand their emotions to protect themselves against stress and to be professional in difficult situations. As the physicians did not want to write down emotions, we used casual self-assessments to be done with little extra effort while documenting (e.g. "How well did the relative(s) understand…" Fig. 2 bottom). The physicians liked these assessments as they took them only little time.

3.4   Case 4: Cooperative Modeling for Non-modelers

In a series of studies, we investigate support cooperative process modeling by people who have not been trained to work with models and thus lack sufficient knowledge on modeling languages and tools (**case 4**, [17, 18]). In one study with five pairs of two participants each, we asked the participants to use a tool originally developed for brainstorming to create a process model (see Fig. 3 left, and [18]). This tool takes textual input of element names and converts it into model elements. We asked the participants to use the tool to write down tasks they perform in a certain process and afterwards adapt and compare their results. Only three participants used models regularly in their work, the others had little experience with models; most of them were familiar with the original brainstorming tool.

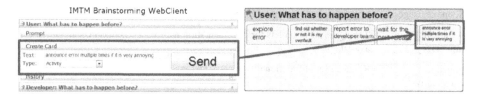

Fig. 3. Using a text input field (left) and translating input into elements of a model (right)

All study participants performed the tasks easily, using text input for adding model elements intuitively and creating reasonable models. They only made a few changes to the models afterwards. This suggests that they had created a rather correct sequence. They even seemed comfortable with explaining the models to each other while comparing them. We could not see significant differences between those who had already been familiar with modeling and those who had little experience with that. This suggests that text input enables modeling for lay modelers, giving gives them a simple subset of the functions needed to create models and fitting their needs.

4   Individual, Group and Situation Adequacy of Access

The dependency of access on users and situations follows from existing work (see section 2). The study details this, showing that individual (e.g. physicians' preferences for different devices to document relative talks in case 3) and group (e.g. the textual description of model elements by lay modelers in case 4) needs are decisive. Table 1 shows factors of access choice such as (individual) usage focus, habits and usage metaphors as well as knowledge and capabilities of individuals or groups.

Table 1. Aspects of individual and group adequacy derived from the case studies

Influence factor	Example from the cases	Level
Usage focus	Logging into an information system vs. integrating its content and functionality in own ways of accessing information (case 1).	Individual
Habits and usage metaphor	Using well-known artifacts and procedures (Pen&Paper) to use a system for ordering and configuring services (case 2).	Group, (individual)
Knowledge and capabilities	Contributing to process models by textual input instead of modeling tool functionality (case 4).	Individual, Group

Adequacy of access also depends on the usage context, e.g. when time is critical like in hospitals (case 3). Table 2 shows factors derived from our cases, including time and scope of usage, availability or immediacy of access and functional complexity.

Table 2. Aspects of situational adequacy derived from the case studies

Influence factor	Example from the cases
Time for usage	Overseeing information in widgets instead of logging into a system (case 1).
Scope of usage	Logging into at system or using it in parallel to other tasks (case 2).
Availability and immediacy	Taking forms or tablets for documentation with you, whereas PCs are not available when needed (case 3).
Functional complexity	Reduced set of functionality such as text input to model elements (case 4).

The factors described above contextualize the influence of users and usage contexts on choosing access. Further work might add or change factors, but the listing given above already provides understanding the adequacy of and choosing proper access.

5   Design Constraints for Access to Cooperation Systems

An analysis of the cases also suggests that access is not a factor driven by technology but make use of it and applying socio-technical patterns of enabling social tasks of individuals. Below, a corresponding list of qualities of access is given:

The Ability to Access Cooperation Support from Different Contexts: In the cases, users wanted to use systems in different contexts – exclusively or in parallel to other tasks (e.g. case 1) and regularly or only in scheduled occasions (case 3). Thus, it is essential to analyze the qualities systems bring into such contexts (e.g. whether they need to be used exclusively). This resembles the context and purpose described by [10] (section 2), but explicitly regards the fit of technology to a certain context.

The Ability to Use Cooperation Support Naturally: Cooperative modeling by text input (case 4) and the support of cooperation with Pen&Paper technology (case 2) show that access depends on the provision of interfaces or devices that users are familiar with – even if these are not common for the tasks and systems involved. This resembles approaches supporting the same interaction with different devices (see [13] and section 2), but adds that choosing access is a matter of transferring known metaphors into the context of using cooperation systems for certain users.

The Ability to Fit Cooperation System Usage into Personal Rhythms: Cases such as using the learning system by widgets (case 1) or collaboratively reflecting in hospitals (case 3) show that access needs to enable individuals to embed system usage into their own rhythms and preferences (filling gaps between tasks, staying within attention spans; see [19] for similar insights) and has to be regarded in design. Mobility is one factor here (case 3), other factors are whether a system or its parts can be used casually (case 3) or which purpose access is used for (case 1, see also [10]).

The Ability to Access Cooperation Systems from Proper Devices: Seniors ordering services with Pen&Paper and physicians needing a tablet and Pen&Paper version of the same tool suggest that access choice needs to regard both the user and the context she is in – for people at the age of 80 and more, mobile phones are not an option and preferences of physicians differ within the group. Choosing an adequate **device** based on these factors needs to be taken into account when access is designed.

While the list given above is non-exhaustive, it already supports designers in choosing access and asking the right questions in design such as whether a system should be used exclusively in a situation or how it fits into work rhythms. However, the list needs to be checked against additional cases and adapted accordingly.

6   Conclusion and Outlook

This paper describes access as a factor for the design of cooperation systems. It extends the existing notion of the term access beyond overcoming limitations and understands it as a vehicle for individuals to use cooperation systems to perform social tasks. The paper builds on a rich base of literature and analyses four cases of small group cooperation support systems, asking which factors influence the choice of access and how access can be designed systematically. Taking into consideration that an analysis of four cases can only be explorative and thus must provide preliminary results, it presents a differentiation of factors describing the influence of users and situation contexts on access choice as well as design qualities to be considered when choosing access. This provides initial answers of the research questions stated above and extends existing insights, putting emphasis on the inherent diversity within user groups and contexts and on the qualities that technologies used for access bring into usage contexts. Further work on the concept will include an analysis of additional cases and design studies applying the insights described above.

References

1. Grudin, J.: CSCW: Time Passed, Tempest, and Time Past. Interactions, 38–40 (2010).
2. Huang, E.M., Harboe, G., Tullio, J., Novak, A., Massey, N., Metcalf, C.J., Romano, G.: Of social television comes home: a field study of communication choices and practices in tv-based text and voice chat. In: Proceedings of the 27th International Conference on Human Factors in Computing Systems, pp. 585–594. ACM, New York (2009)
3. Zimmermann, G., Vanderheiden, G.: Accessible design and testing in the application development process: considerations for an integrated approach. Universal Access in the Information Society 7, 117–128 (2008)
4. Kaptelinin, V., Nardi, B.A.: Acting with Technology. In: Activity Theory and Interaction Design. MIT Press, Cambridge (2005)
5. Giddens, A.: Elements of the Theory of Structuration. In: The Constitution of society: Outline of the Theory of Structuration, Berkeley, CA, pp. 1–40 (1984)
6. Orlikowski, W.J.: Using Technology and Constituting Structures: A Practice Lens for Studying Technology in Organizations. Organization Science 11, 404–428 (2000)

7. Dourish, P.: Seeking a foundation for context-aware computing. Human-Computer Interaction 16, 229–241 (2001)
8. Driskell, J.E., Radtke, P.H., Salas, E.: Virtual Teams: Effects of Technological Mediation on Team Performance. Group Dynamics: Theory, Research, and Practice 7, 297 (2003)
9. Löwgren, J.: Articulating the use qualities of digital designs. Aesthetic computing. pp. 383–403. The MIT Press (2006)
10. Jung, H., Stolterman, E., Ryan, W., Thompson, T., Siegel, M.: Toward a framework for ecologies of artifacts: how are digital artifacts interconnected within a personal life? In: Proceedings of the 5th Nordic Conference on Human-Computer Interaction: Building Bridges, pp. 201–210 (2008)
11. Dennis, A.R., Valacich, J.S.: Rethinking media richness: Towards a theory of media synchronicity. In: HICSS-32, Proceedings of the 32nd Annual Hawaii International Conference on System Sciences, pp. 10–19 (1999)
12. Davis, F.D.: User acceptance of information technology: system characteristics, user perceptions and behavioral impacts. International Journal of Man-Machine Studies 38, 475–487 (1993)
13. Nylander, S., Bylund, M., Waern, A.: Ubiquitous service access through adapted user interfaces on multiple devices. Personal and Ubiquitous Computing 9, 123–133 (2005)
14. Prilla, M., Ritterskamp, C.: The Interplay of Web 2.0 and Collaboration Support Systems: Leveraging Synergies. In: Randall, D., Salembier, P. (eds.) From CSCW to Web 2.0: European Developments in Collaborative Design. Selected Papers from COOP 2008. Springer (2010)
15. Loser, K.-U., Nolte, A., Prilla, M., Skrotzki, R., Herrmann, T.: A Drifting Ambient Assisted Living Project. In: Viscusi, G., Campagnolo, G.M., Curzi, Y. (eds.) Phenomenology, Organizational Politics and IT Design: The Social Study of Information Systems. IGI Global (2012)
16. Prilla, M., Degeling, M., Herrmann, T.: Collaborative Reflection at Work: Supporting Informal Learning at a Healthcare Workplace. In: Proceedings of the ACM International Conference on Supporting Group Work, GROUP 2012 (forthcoming, 2012)
17. Herrmann, T., Nolte, A., Prilla, M.: The Integration of Awareness, Creativity Support and Collaborative Process Modeling in Colocated Meetings. International Journal on Computer-Supported Cooperative Work (IJCSCW). Special Issue on Awareness (2012)
18. Prilla, M., Nolte, A.: Integrating Ordinary Users into Process Management: Towards Implementing Bottom-Up, People-Centric BPM. In: Bider, I., Halpin, T., Krogstie, J., Nurcan, S., Proper, E., Schmidt, R., Soffer, P., Wrycza, S. (eds.) EMMSAD 2012 and BPMDS 2012. LNBIP, vol. 113, pp. 182–194. Springer, Heidelberg (2012)
19. Höök, K.: Transferring qualities from horseback riding to design. In: Proceedings of the 6th Nordic Conference on Human-Computer Interaction: Extending Boundaries, pp. 226–235 (2010)

Characterizing Key Developers:
A Case Study with Apache Ant

Gustavo A. Oliva[1], Francisco W. Santana[2], Kleverton C.M. de Oliveira[2],
Cleidson R.B. de Souza[2,3], Marco A. Gerosa[1]

[1] Department of Computer Science, University of São Paulo (USP), Brazil
{goliva,gerosa}@ime.usp.br
[2] Computing Department, Federal University of Pará (UFPA), Brazil
[3] Vale Technological Institute – Sustainable Development (ITV – DS), Brazil
{wertherjr,kleverton.macedo}@gmail.com,
cleidson.desouza@acm.org

Abstract. The software architecture of a software system and the coordination
efforts necessary to create such system are intrinsically related. Making changes
to components that a large number of other components rely on, the technical
core, is usually difficult due to the complexity of the coordination of all involved developers. However, a distinct group of developers effectively help
evolving the technical core of software projects. This group of developers is
called key developers. In this paper we describe a case study involving the
Apache Ant project aimed to identify and characterize key developers in terms
of their volume of contribution and social participation. Our results indicated
that only 25% of the developers may be considered as key developers. Results
also showed that key developers are often active in the developers' mailing list
and often fulfilled the coordination requirements that emerged from their development tasks. Finally, we observed that the set of key developers was indistinguishable from the set of top contributors. We expect that this characterization
enables further exploration over contribution patterns and the establishment of
profiles of FLOSS key developers.

Keywords: software architecture, collaboration, socio-technical analysis, mining software repositories, case study.

1   Introduction

In the 60s', Conway [6] suggested that the relationship between the architecture of a
software system and the structure of the organization developing this software is ho-
momorphic – the Conway's Law. Similarly, Parnas [17] suggested an approach, the
information hiding principle, to structure the software architecture in such a way to
reduce coordination needs among developers. Recently, these theoretical proposals
have been corroborated by several qualitative [25, 10, 24] and quantitative [5, 4]
studies.

V. Herskovic et al. (Eds.): CRIWG 2012, LNCS 7493, pp. 97–112, 2012.
© Springer-Verlag Berlin Heidelberg 2012

These results basically suggest that the structure of a software system influences and is influenced by the communication and coordination efforts of the developers developing such system. Furthermore, the coordination necessary to evolve highly interconnected software components is usually greater than the effort required to evolve independent components. This seems to be the case even when well-defined APIs are used [24]. In fact, despite the rhetoric about openness, access to the technical core of a software project (the set of the most important software components on which other components rely on) is limited [22]. Apart from that, we cannot say much more about the group of developers that help evolve the technical core. Are these *key developers* the ones that communicate more to other developers in the mailing-list? Are they the ones in the core of the coordination requirements network? Are they the ones that have higher socio-technical congruence [5] when considering the mailing-list network (social activities that actually occurred) and the coordination requirements network (social activities that should have taken place)? Are they also the top committers? In a long term perspective, a better characterization of key developers should help researchers understand the process a developer undergoes in order to become a key developer.

In this paper, we describe a case study conducted with the open source project Apache Ant[1] in order to investigate the characteristics of its *key developers*, i.e., the set of developers that work on the technical core of this project. Firstly, we designed and applied an appropriate method to evaluate how limited the number of key developers is. Afterwards, we investigated whether these developers (i) were central in the mailing-list network, (ii) were central in the coordination requirements network, (iii) had a higher congruence when considering these two social networks [5], or (iv) were just the top committers. Our results indicated that only 25% of the developers were classified as key developers. Results also showed that key developers were active in the developers' mailing list and often fulfilled the coordination requirements that emerged from their development tasks (high socio-technical congruence). Finally, we observed that the set of key developers was indistinguishable from the set of top contributors.

The rest of this paper is organized as follows. In Section 2, we present our research questions. In Section 3, we present related work. Section 4 then describes the research method, including the supporting tools we used. Our results are presented in Section 5. After that, Section 6 presents a discussion of our results and describes the threats to the validity of this study. Finally, in Section 7, we state our conclusions and plans for future work.

2   Characterizing Key Developers

The relationship between the architecture of a software system and the coordination required to evolve such a system is long recognized by researchers and practitioners. For instance, the performance of software developers is related to how well they align

---

[1]   http://ant.apache.org

their coordination efforts with the existing technical dependencies in the software architecture, both at the team level [25] and at the individual level [4]. Indeed, misalignment between these aspects is seen as a possible explanation for breakdowns in software development projects [2]. In other words, the relationship between software architecture and coordination suggests that the coordination effort necessary to develop highly interconnected software components is usually higher than to develop independent components. This is true even when well-defined interfaces are used among software components [24].

In any software system, there are components that are regarded as more important than others. Such components constitute the technical core of a project, i.e. the set of the most important software components on which lots of other components rely on. In this paper, we call *key developers* the set of developers who help evolve the technical core of a software system. Given the existing relationship between software architecture and coordination, we expect the access to the technical core of a software project to be limited. This aspect has already been observed in previous studies of open source projects [22]. In other words, we expected a limited number of key developers. The reason is twofold: (i) the technical core is naturally important (if someone "breaks" a core component, then several other components are likely to be affected) and (ii) the complexity of the coordination necessary to make changes to the core is high. This leads to our first research question:

RQ 1: How limited is the number of key developers in a software project?

Social interaction within software development is acknowledged as an important aspect in software projects and thus has been the subject of a series of studies [6, 17, 10, 23]. Different social processes (e.g., development of a shared understanding of the system architecture, conflict resolution, and leadership establishment) are necessary for successful projects. These social processes often involve key developers differently from the rest [8], we believe that a better characterization of such developers would be beneficial to researchers interested in collaborative software development. The investigation of key developers seems especially suitable in the context of free/libre open source software development (FLOSS development) and global software development (GSD), where social interaction data is usually available in software repositories and in the project's website. This leads us to our second research question.

RQ 2: How distinct is the participation of key developers in terms of communication and coordination?

While conducting two case studies involving the Apache Server and the Mozilla web browser respectively, Mockus *et al.* [13] proposed the following hypothesis: "*open source developments will have a core of developers who control the code base, and will create approximately 80% or more of the new functionality. If this core group uses only informal, ad hoc means of coordinating their work, it will be no larger than 10-15 people.*" As our goal in this study involves characterizing key developers, we also intend to verify whether a relaxed version of such hypothesis also holds for the Apache Ant project. More specifically, instead of looking for added functionality, we will just analyze the number of modifications made by each developer. We operationalize that by identifying the group of *top contributors*, i.e. the set of developers who performed the highest number of modifications (commits) to the project. We thus state our last research question as follows.

RQ 3: What is the contribution volume of key developers?

3   Related Work

A number of previous studies have investigated the participation of open source developers regarding their "status" position within the community. For instance, Crowston *et al.* [8] examined how the group of core developers can be empirically distinguished. The authors investigated three specific approaches, namely (i) the named list of developers, (ii) the most frequent contributors, and (iii) a social network analysis of the developers' interaction pattern. By applying these three approaches to the interactions around bug fixing for 116 SourceForge projects, the authors concluded that each approach identify different individuals as core developers. However, as in our paper, the results suggest that the group of core developers in FLOSS projects corresponds to only a small fraction of the total number of contributors. In another example, Terceiro *et al.* [26] investigated the relationship between code structural complexity and the participation level of developers (dichotomized as core and peripheral). By relying on previous studies of Robles *et al.* [18, 19], the authors split the entire studied period in 20 periods of equal duration, and for each period, they considered the 20% top committers to be the core team. They found out that core developers make changes to the source code without introducing as much structural complexity as the peripheral developers. Moreover, core developers also remove more structural complexity than peripheral developers.

Other studies have focused on investigating the characteristics and behavior of software developers from a social network analysis (SNA) perspective. De Souza *et al.* [22] investigated the ways in which development processes are somehow inscribed into software artifacts. The authors hypothesized that when developers shift from the periphery to the core of the code authorship social network, a distinct phenomenon occurs. Developers initially contribute code that performs some functionality by calling others' code and, as these developers become more important, their code start to be called by other developers. De Souza and colleagues showed a periphery to core shift within the MegaMek project, and a core to periphery shift (opposite effect) within the Apache Ant project. In another study, Oezbek *et al.* [16] investigated the patterns of interaction among the core and peripheral sets of developers in order to check the validity of the "onion model" [14]. After building social networks based on mailing lists data from 11 FLOSS projects of different domains, the authors observed that the core holds a disproportionally large share of communication with the periphery. They also state that members of the core not only show a particular intense participation, but also appears to have a qualitatively different role as well. However, such hypothesis remains to be investigated. The authors also conclude that the transition of individual mailing-list participants towards ever higher participation is qualitatively discontinuous.

4   Research Method

In order to answer the research questions defined previously, we decided to adopt a case study as our research method. A case study is a well-established empirical method aimed at investigating contemporary phenomena in their natural context [28]. More specifically, we conducted a *descriptive case study with retrospective data collection* [20]. In this case study we sought to portray the characteristics of key developers by leveraging the project's available stored data. In contrast to embedded case studies, where multiple units of analysis are studied within a case, our case study is essentially holistic, i.e. the case is studied "as a whole." In a nutshell, we focused on a particular open source project and gathered different types of information from it.

In the next subsections, we present the case study design and planning. We present the rationale for choosing the case, the supporting tools we used, and the main steps we followed.

4.1   The Case

For the case study, we needed a software project that satisfied the following requirements: (i) a software project hosted on a Subversion (SVN) repository with anonymous read access; (ii) availability of information about the development activities (change logs and communication records) during a release interval, and (iii) a number of active developers greater than 15. The first requirement exists due to constraints on the tools at our disposal. The second requirement was raised because we need development information to generate the social networks and compute volume contribution. Furthermore, we will focus our analysis on a specific release interval so as to minimize influencing factors. Finally, the third requirement came up because we need sufficient social data to answer our research questions. Hence, we decided to focus on non-small development teams: Levine and Moreland [11] defined small teams as groups of 5 to 15 individuals.

After inspecting a series of open source projects, we decided to analyze Apache Ant: it is hosted on Subversion, information about development activities is available, and 16 developers contributed to it during the studied release interval. More precisely, we investigated Apache Ant Core, which is the main Ant module. We considered a development period that ranges from release 1.6 (December 19th, 2003) until release 1.7 (December 13th, 2006). In such period, a total of 2053 commits were made by a group of 16 active developers. Apache Ant is hosted by the Apache Software Foundation and is one of the most popular open source tools for automating software build processes.

4.2   Supporting Tools

Empirical studies that mine software repositories usually require extensive tool support due to the large amount and complexity of the data to be collected, processed, and analyzed [22]. Given the different data sources required in this study, we employed a variety of tools: XFlow [21], JDX, Jung[2], and OSSNetwork [1].

XFlow. XFlow is an extensible open source tool we developed whose main goal is to support empirical software evolution analyses by considering both social and technical aspects. By bringing together these two views, the tool aims to support exploratory and descriptive case studies that call for a deeper understanding of software evolution aspects. In this study, we employed XFlow to calculate the coordination requirements network [5].

JDX. Java Dependency eXtractor is a Java library we developed to extract dependencies and compute the call-graph from Java code. The library relies on the robust Java Development Tools Core (JDT Core)[3] library, which is the Eclipse IDE incremental compiler. As a desirable consequence, JDX is able to handle Java source code in its plain form. This facilitates studies that involve processing large amounts of code mined from version control systems.

OSSNetwork. OSSNetwork is a tool we developed that (i) retrieves data from software repositories (forums, mailing lists, issue tracking systems, and chats) by parsing HTML information and (ii) generates different social networks, thus supporting the analysis of social aspects of software development. We used OSSNetwork to compute the communication network from the developers' mailing list.

Jung. Java Universal Network/Graph Framework is a Java library that provides a common and extendible language for modeling, analyzing, and visualizing data that can be represented as a graph or network. We used Jung to compute network properties, such as the eigenvector centrality of nodes (as will be detailed in the following subsection).

4.3   Main Steps

In order to answer our research questions we mined Apache Ant's development repositories, namely the version control system (Subversion) and the developers' mailing list. This whole process was divided into three main steps:

I) Identifying Key Developers. The varying complexity of software system modules requires an equally varied amount of knowledge from developers in order to complete their tasks. As we are interested in characterizing key developers, our first step was to discover which developers actually worked on the core files of the Apache Ant project. In other words, this investigation requires finding both the core of the technical network and the particular developers that worked on such core. Hence, for each Subversion revision embedded in the studied development period, we did the following sub-steps:

---

[2] http://jung.sourceforge.net/
[3] http://www.eclipse.org/jdt/core/

a) Generate the project's technical network. We calculated the project's static call-graph using JDX. According to Wikipedia, a static call-graph is a directed graph that represents calling relationships between subroutines in a computer program. In our context, each node represents a method and each edge (f, g) indicates that a method f calls a method g (including constructor invocations). After obtaining the call-graph, we clustered the method nodes belonging to the same compilation unit. We thus obtained a new graph in which the nodes represent the compilation units and the edges represent their calling relationship. We considered such graph to be a suitable representation of the project's technical network.

b) Finding the core of the technical network. We used the eigenvector centrality measure to find the core of the network produced in the prior step. Such measure embodies the notion that a node's importance in a network is increased by having connections to other vertices that are themselves important [15]. Indeed, we believe that a compilation unit becomes important by having connections to other compilation units that are themselves important. We calculated the centrality of each node of the network and then we performed a quartile analysis to identify the network's core. The nodes that had a centrality score equal to or larger than the third quartile (Q3) were deemed as core.

c) Computing commit coreness. In order to differentiate developers' contributions, we conceived a measure for computing the *commit coreness*. This measure is calculated based on the number of modified core artifacts, thus enabling us check whether a developer actually contributed to the technical core or just made peripheral changes:

$$Coreness(commit) = \frac{Number\ of\ core\ files\ in\ the\ commit}{Total\ number\ of\ files\ in\ the\ commit}$$

When *commit coreness* was greater than or equal to 0.5, we considered it to be a core commit. In fact, when a core commit was detected, we considered that its author made a modification to the technical core of the system.

II) Social Network Analysis. Given the list of key developers obtained from the previous step, we investigated whether they (i) belonged to the core of the communication network (mailing list activity), (ii) belonged to the core of the coordination requirements network, (iii) had a high congruence when considering these two networks, or (iv) were top committers. In the following, we briefly describe how we evaluated these four scenarios respectively.

a) Developers in the core of the project's communication network. We collected data from the developers' mailing list using the OSSNetwork tool and built a communication network in the form of an undirected graph. Links were established by detecting developers that contributed to a same mail thread (including the original email sender). For instance, if developer a sends an email, and developers b and c reply to it, then links among all these developers are added to the communication network. Analogously to what was done for the technical network (step I.b), we identified the developers that were in the core of this network by employing the eigenvector centrality measure and doing a quartile analysis.

b) Developers in the core of the coordination requirements network. We generated the project's coordination requirements network using the method proposed by Cataldo *et al.* [5], which relies on the concept of evolutionary dependencies [9]. Such dependencies consist of implicit relationships that are established between software artifacts as they are frequently changed together. This network depicts the set of individuals a developer should coordinate his/her work with (or at least be aware of), since their work tasks share a certain level of interdependency [5, 7]. With the coordination requirements network in hands, we again used eigenvector centrality and a quartile analysis to identify the core of the social network, just as in the previous scenario.

c) Congruence between these two networks. Inspired by the measure of congruence defined Cataldo *et al.* [5], we computed the proportion of social activity that actually occurred (given by the communication network extracted from the mailing list) relative to the social activity that should have taken place (given by the coordination requirements network extracted from the evolutionary dependencies) for each developer. Congruence values thus range between 0 and 1. Such approach for measuring congruence builds on the idea of "fit" from the organizational theory literature [3]. We performed a quartile analysis and the congruence values that were equal to or larger than the third quartile (Q3) were deemed as high.

d) Top contributors. We intend to check whether a small number of developers are responsible for most part of the modifications made to the software system. By using XFlow we computed the top contributors of the Apache Ant project during the studied timeframe, i.e. those developers that made most part of the commits. More precisely, we determined the top committers by analyzing the distribution of *commits per developer*.

III) Comparative Analyses. The final step involved comparing the set of key developers obtained in step I.c with the developers that we identified in the steps II.a, II.b, II.c, and II.d. The results are described in the following section.

5   Results

After collecting the project data by following the aforementioned methods, we divided the results into three groups: the identification of key developers, the analysis of the project's social networks, and the identification of top contributors. In the next subsections we will thoroughly discuss each group of results.

5.1   Identification of Key Developers

As mentioned before, we used JDX to compute the technical network of the codebase corresponding to each Subversion revision of Apache Ant. We then calculated the core of such network and decided whether each revision actually involved a change to the technical core. After that, we calculated the number of core modifications made by each developer. In Table 1 we depict the results we obtained:

Table 1. Developers and Associated Number of Core Modifications to the System

Developer	Number of Core Commits	Delta
ddevienne	0	0
scohen	0	0
umagesh	1	1
conor	1	0
alexeys	2	1
bruce	3	1
jhm	3	0
sbailliez	4	1
kevj	14	10
antoine	25	11
jglick	27	2
jkf	40	13
stevel	77	37
bodewig	118	41
mbenson	172	54
peterreilly	178	6

We sorted the developers according to number of core commits they performed. The thrid column of the table (delta) shows the difference between the number of commits of a developer and his predecessor. The data in this column indicates a first major shift from *jkf* to *stevel* (37). In fact, we notice that approximately 82% of the core commits are performed by a specific group of four developers: *stevel*, *bodewig*, *mbenson*, and *peterreilly*. Therefore, we considered those to be the key developers of Apache Ant during the studied release period.

5.2   The Different Social Networks

In this section, we present the two different social networks we obtained, as well as the measure of congruence for each developer in the Apache Ant project during the studied period.

The Core of the Communication Network. We used OSSNetwork to compute the communication network of the project. Fig. 1 depicts the result we obtained in the form of a graph, in which vertices represents project's developers and edges maps the existence of mail exchanged between two linked vertices.

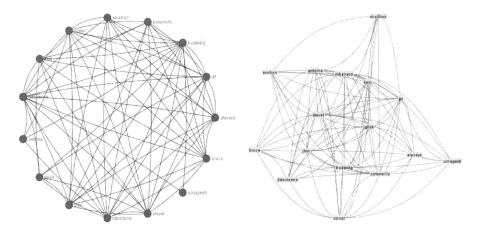

Fig. 1. Communication Network of Apache Ant

Fig. 2. Coordination Requirements Network of Apache Ant

After that, we employed the Eigenvector centrality measure and the quartile analysis to obtain the set of developers in the core of this network. The results indicated that four individuals are in the core: *bodewig*, *mbenson*, *stevel,* and *jkf.*

The Core of the Coordination Requirements Network. We used XFlow to apply the method proposed by Cataldo *et al.* [5] to calculate the coordination requirements network. Fig. 2 depicts XFlow's graph view of the coordination requirements, where each vertex represents a developer and each edge maps two developers that are likely to coordinate their efforts because the artifacts they are changing are interdependent. After that, analogously to the previous case, we calculated the eigenvector centrality and performed a quartile analysis to obtain the set of developers belonging to the core of this network. The results indicate that a large number of individuals are in the core: *peterreilly, bodewig, mbenson, stevel, jkf, jglick, antoine, alexeys, jhm, sbailliez, conor, bruce, kevj,* and *ddevienne*. Only two developers are not in this list, namely *umagesh* and *scohen.*

Congruence of the Networks. We computed the socio-technical congruence of these two networks for each developer. Fig. 3 depicts the results we obtained. The data shows that the interval of congruence values is large (ranging from 90% to 0%). In a similar fashion to the previous cases, we performed a quartile analysis in order to identify developers with higher congruence. The results we obtained pointed out to four individuals: *ddevienne, bodewig, kevj,* and *stevel.*

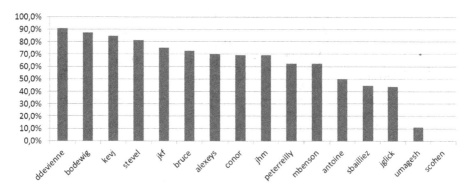

Fig. 3. Socio-technical congruence of the developers

5.3   Top Contributors

We used XFlow and calculated the top contributors of the Apache Ant project during the analyzed period. Fig. 4 depicts the cumulative percentage of the number of commits. According to the data, 4 developers (25% of them) were responsible for 81% of the commits.

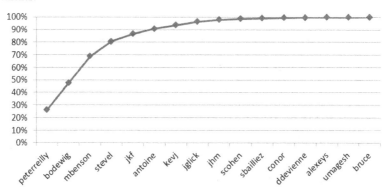

Fig. 4. Cumulative percentage of the number of commits

Therefore, we conclude that the relaxed version of Mockus' hypothesis we defined indeed holds for the Ant project, as most part of the modifications (commits) are made by a small group of developers.

6   Discussion

We start the discussion by illustrating the intersection between the set of key developers and those that (i) are in core of the communication network, (ii) are in the core of the coordination requirements network, (iii) have high socio-technical congruence, and (iv) are top contributors. These results are presented in Table 2.

Table 2. Characterizing key developers

Developer	Key Developer	Core of Communication Network	Core of Coordination Requirements Network	High Congruence	Top Contributors
peterreilly	✓		✓		✓
bodewig	✓	✓	✓	✓	✓
mbenson	✓	✓	✓		✓
stevel	✓	✓	✓	✓	✓
jkf		✓	✓		
jglick			✓		
antoine			✓		
alexeys			✓		
jhm			✓		
sbailliez			✓		
conor		✓	✓		
bruce			✓		
kevj			✓	✓	
ddevienne		✓	✓	✓	
umagesh					
scohen					

Only four key developers were identified, namely: *peterreilly*, *bodewig*, *mbenson*, and *stevel*. Three of these key developers also belonged to the core of the communication network (although such core includes three other developers). This provides evidence that most key developers were also very active in the developers' mailing list during the analyzed period. In relation to the core of the coordination requirements network, all key developers belonged to it. This was somehow expected, since the core of the coordination requirements included 14 of the 16 developers. We think that such core was large due to the inclusive nature of the algorithm used for computing this network: no filters were applied to the evolutionary dependencies, which means that even dependencies between components that occurred only once in the analyzed period are taken into account. We also computed the socio-technical congruence of these two networks for each developer and we noticed that two of the key developers had high congruence. On the other hand, the results also suggest that although *kevj* and *ddeviene* were very communicative (in the sense that they communicated with almost everyone they were required to), they did not work on the technical core very often. Interestingly, the sets of key developers and top contributors are identical (perfect correlation). In fact, by taking a closer look at the volume contribution data, we can see that the set of key developers also heavily contributed to the peripheral areas of the technical network. Finally, only two developers did not show up in any of the considered cases, namely: *umagesh* and *scohen*.

We now answer our research questions in light of the results we obtained. The first question concerned how limited the number of key developers is. As we presented, only four developers (25%) were responsible for approximately 82% of the core modifications. This corroborates our initial expectation that only a few developers would be responsible for making changes to the core. The second question concerned how key developers coordinated their efforts and communicated. After analyzing the developers' mailing list, the coordination requirements network, and the congruence between these two networks (the socio-technical congruence), we noticed that two of the key developers (*bodewig* and *stevel*) were in the core of both networks and also presented a high congruence. The developer *mbenson* was solely in the core of both networks. The developer *peterreilly*, in turn, only appeared in the core of the coordination requirements network. In general, this provides evidence that key developers were often very active in the mailing list (except for *peterreilly*, who was not very active within the project's mailing list). Given the strong connection between software architecture and coordination, we believe that such social interaction help developers coordinate their tasks and keep themselves aware of changes made to the software system. Our third and last research question concerned the contribution volume of key developers. The results showed that key developers were also the ones that contributed the most to the project.

6.1   Threats to Validity

There are some factors that may have influenced the validity of our study.

Construct Validity. Firstly, a common practice in FLOSS development concerns the submission of patches by non-developers interested in helping a particular software project. As these users do not have permission to commit their fixes on the projects' version control system, their contributions are often committed by one of the regular project developers. As a result, this may have introduced some noise in the data used to calculate key developers. Secondly, the webcrawler algorithm employed by OSS-Netwok to parse mailing list data (from HTML pages) makes use of semi-structured webpages as source of information, which is clearly subject to problems due to the lack of rigid rules for participation and participants' identification in the mailing lists. Thirdly, the adoption of eigenvector centrality metric to define core sets on networks might affect our findings. We believe that this measure captures a behavior that seems adequate to our analysis, but other approaches (e.g. k-core or islands) could provide different results. Finally, other thresholds could have been used to determine whether a modification (commit) is core or not.

Internal Validity. Our empirical evidences cover only a single release of the Apache Ant, and it is thus possible that we missed empirical evidence that could be found in other releases of the same project. A more extensive study should be conducted in order to further investigate key developers' characteristics in terms of their social interaction and contributions.

External Validity. Since we studied a single project, we cannot state that these results would remain valid for other projects. In fact, threats to the generalizability of this study are given by the very nature of the employed research design. McGrath [12] states that research methods can be evaluated on three dimensions (generalizability, realism, and precision) and he argues that no method is able to satisfy all dimensions at the same time. In particular, case studies naturally maximize realism, but seldom satisfy generalizability (since they involve a small number of non-randomly selected situations) or precision (because there is a low level of control over influencing factors). Hence, we leverage the realism of our results and conclusions.

7   Conclusion and Future Work

In this paper, we presented a descriptive case study involving Apache Ant. Our goal was to characterize *key developers*, i.e. those developers that effectively evolve the technical core of the project. The reason for studying them is that the access to the technical core of a software project is often restricted to a few developers. In particular, we were interested in answering three research questions that involved investigating (i) *how limited the number of key developers is*, (ii) *how distinct the participation of key developers is (in terms of communication and coordination)*, and (iii) *the contribution volume of key developers*. Our results indicated that only 25% of the developers were classified as key developers. We also showed that key developers were often active in the developers' mailing list and often fulfilled the coordination requirements that emerged from their development tasks. Finally, we noticed that the set of key developers was identical to the set of top contributors.

Our expectations with our findings are that in a long term perspective better characterizing key developers should help researchers understand the process a developer undergoes in order to become a key developer. As these key developers play a crucial role in the project, properly characterizing and identifying them is important in order to better understand the various social processes that often occur within software development. Furthermore, although prior research has tried to understand the process of core-periphery migration on FLOSS projects, the identification of the set of core developers has always been a difficult task that is mostly performed using purely visual methods, which end up posing threats to the validity of these studies and emphasizing the need for more accurate methods.

As future work, we believe that applying our research method to different FLOSS and commercial projects will help verify whether key developers characteristics are similar to those we reported.

Acknowledgements. We thank Steve Abrams for his insightful contribution to the design of the research method, more specifically the usage of eigenvector centrality. Gustavo Oliva receives individual grant from the CHOReOS EC FP7 project. Marco Gerosa receives individual grant from CNPq. Cleidson de Souza was supported by FAPESPA through "Edital Universal N° 003/2008".

References

[1] Balieiro, M., de Júnior, S., de Souza, C.: Facilitating Social Network Studies of Floss Using the Ossnetwork Environment. In: Russo, B., Damiani, E., Hissam, S., Lundell, B., Succi, G. (eds.) Open Source Development, Communities and Quality. IFIP, vol. 275, pp. 343–350. Springer, Boston (2008), http://dx.doi.org/10.1007/978-0-387-09684-1_31, 10.1007/978-0-387-09684-1_31

[2] Bass, M., Mikulovic, V., Bass, L., Herbsleb, J., Cataldo, M.: Architectural misalignment: An experience report. In: Proceedings of the Sixth Working IEEE/IFIP Conference on Software Architecture, WICSA 2007, p. 17. IEEE Computer Society, Washington, DC (2007), http://dx.doi.org/10.1109/WICSA.2007.12

[3] Burton, R.M., Obel, B.: Strategic Organizational Diagnosis and Design: The Dynamics of Fit, 3rd edn. Information and Organization Design Series. Springer (2003)

[4] Cataldo, M.: Dependencies in geographically distributed software development: overcoming the limits of modularity. Ph.D. thesis, Pittsburgh, PA, USA (2007), aAI3292617

[5] Cataldo, M., Wagstrom, P., Herbsleb, J.D., Carley, K.M.: Identification of coordina-tion requirements: implications for the design of collaboration and awareness tools. In: Hinds, P.J., Martin, D. (eds.) CSCW, pp. 353–362. ACM (2006), http://doi.acm.org/10.1145/1180875.1180929

[6] Conway, M.: How do committees invent. Datamation 14(4), 28–31 (1968)

[7] Costa, J.M., Cataldo, M., de Souza, C.R.: The scale and evolution of coordination needs in large-scale distributed projects: implications for the future generation of collabo-rative tools. In: Proc. of the 2011 Annual Conference on Human Factors in Computing Systems, CHI 2011, pp. 3151–3160. ACM (2011), http://doi.acm.org/10.1145/1978942.1979409

[8] Crowston, K., Wei, K., Li, Q., Howison, J.: Core and periphery in free/libre and open source software team communications. In: Proceedings of the 39th Annual Hawaii International Conference on System Sciences, HICSS 2006, p. 118. IEEE Computer Society, Washington, DC (2006), http://dx.doi.org/10.1109/HICSS.2006.101

[9] Gall, H., Hajek, K., Jazayeri, M.: Detection of logical coupling based on product re-lease history. In: Proceedings of the International Conference on Software Maintenance, ICSM 1998, p. 190. IEEE Computer Society, Washington, DC (1998), http://dl.acm.org/citation.cfm?id=850947.853338

[10] Grinter, R.E.: Systems architecture: product designing and social engineering. SIGSOFT Softw. Eng. Notes 24(2), 11–18 (1999), http://doi.acm.org/10.1145/295666.295668

[11] Levine, J.M., Moreland, R.L.: Progress in small group research. Annual Review of Psychology 41(1), 585–634 (1990), http://arjournals.annualreviews.org/doi/abs/10.1146/annurev.ps.41.020190.003101

[12] McGrath, J.E.: Dilemmatics: The study of research choices and dilemmas. American Behavioral Scientist 25(2), 179–210 (1981), http://abs.sagepub.com/cgi/doi/10.1177/000276428102500205

[13] Mockus, A., Fielding, R.T., Herbsleb, J.D.: Two case studies of open source software development: Apache and mozilla. ACM Trans. Softw. Eng. Methodol. 11, 309–346 (2002), http://doi.acm.org/10.1145/567793.567795

[14] Nakakoji, K., Yamamoto, Y., Nishinaka, Y., Kishida, K., Ye, Y.: Evolution patterns of open-source software systems and communities. In: Proceedings of the International Workshop on Principles of Software Evolution, IWPSE 2002, pp. 76–85. ACM, New York (2002), http://doi.acm.org/10.1145/512035.512055

[15] Newman, M.: Networks: An Introduction, 1st edn. Oxford University Press (2010)

[16] Oezbek, C., Prechelt, L., Thiel, F.: The onion has cancer: some social network analysis visualizations of open source project communication. In: Proceedings of the 3rd International Workshop on Emerging Trends in Free/Libre/Open Source Software Research and Development, FLOSS 2010, pp. 5–10. ACM, New York (2010), http://doi.acm.org/10.1145/1833272.1833274

[17] Parnas, D.L.: On the criteria to be used in decomposing systems into modules. Commun. ACM 15, 1053–1058 (1972), http://doi.acm.org/10.1145/361598.361623

[18] Robles, G., Gonzalez-Barahona, J.: Contributor Turnover in Libre Software Projects. In: Damiani, E., Fitzgerald, B., Scacchi, W., Scotto, M., Succi, G. (eds.) Open Source Systems. IFIP, vol. 203, pp. 273–286. Springer, Boston (2006), http://dx.doi.org/10.1007/0-387-34226-5_28,10.1007/0-387-34226-5_28

[19] Robles, G., Gonzalez-Barahona, J.M., Herraiz, I.: Evolution of the core team of developers in libre software projects. In: Proceedings of the 2009 6th IEEE International Working Conference on Mining Software Repositories, MSR 2009, pp. 167–170. IEEE Computer Society, Washington, DC (2009), http://dx.doi.org/10.1109/MSR.2009.5069497

[20] Robson, C.: Real World Research, 2nd edn. John Wiley & Sons (2002)

[21] Santana, F., Oliva, G., de Souza, C.R.B., Gerosa, M.A.: Xflow: An extensible tool for empirical analysis of software systems evolution. In: Proceedings of the VIII Experimental Software Engineering Latin American Workshop, ESELAW 2011 (2011)

[22] de Souza, C., Froehlich, J., Dourish, P.: Seeking the source: software source code as a social and technical artifact. In: Proc. of the 2005 International ACM SIGGROUP Conference on Supporting Group Work, GROUP 2005, pp. 197–206. ACM (2005), http://doi.acm.org/10.1145/1099203.1099239

[23] de Souza, C.R., Quirk, S., Trainer, E., Redmiles, D.F.: Supporting collaborative software development through the visualization of socio-technical dependencies. In: Proc. of the 2007 International ACM Conference on Supporting Group Work, GROUP 2007, pp. 147–156. ACM (2007), http://doi.acm.org/10.1145/1316624.1316646

[24] Souza, C.R., Redmiles, D.F.: On the roles of apis in the coordination of collaborative software development. Comput. Supported Coop. Work 18(5-6), 445–475 (2009), http://dx.doi.org/10.1007/s10606-009-9101-3

[25] Staudenmayer, N.A.: Managing multiple interdependencies in large scale software development projects. Ph.D. thesis, Massachusetts Institute of Technology (1997)

[26] Terceiro, A., Rios, L.R., Chavez, C.: An empirical study on the structural complexity introduced by core and peripheral developers in free software projects. In: Proceedings of the 2010 Brazilian Symposium on Software Engineering, SBES 2010. IEEE Computer Society, Washington, DC (2010), http://dx.doi.org/10.1109SBES.2010.26

[27] Valetto, G., Helander, M., Ehrlich, K., Chulani, S., Wegman, M., Williams, C.: Using software repositories to investigate socio-technical congruence in development projects. In: Proceedings of the Fourth International Workshop on Mining Software Repositories, MSR 2007, p. 25. IEEE Computer Society, Washington, DC (2007), http://dx.doi.org/10.1109/MSR.2007.33

[28] Yin, R.K.: Case study research: Design and methods, 3rd edn. Sage Publications (2003)

An Exploratory Study on Collaboration Understanding in Software Development Social Networks

Andréa M. Magdaleno[1], Renata M. Araujo[2], and Cláudia M.L. Werner[1]

[1] UFRJ – Federal University of Rio de Janeiro
COPPE – Systems Engineering and Computer Science Department
21945-970, Rio de Janeiro, RJ, Brazil, P.O. Box 68511
[2] Graduate Program in Information Systems (PPGI) – UNIRIO
{andrea,werner}@cos.ufrj.br, renata.araujo@uniriotec.br

Abstract. Collaboration is important for productivity, quality, and knowledge sharing in software development. In this context, the use of social networks analysis can help to track the level of collaboration in a development project. In this work, an exploratory study was conducted, in the context of free/open source software, using EvolTrack-SocialNetwork tool, to investigate collaboration in software teams. The preliminary results indicate a potential to increase one's ability to understand the course that the collaboration is taking.

Keywords: Collaboration, social network, software development.

1   Introduction

Collaboration has been indicated as an important factor for achieving goals in productivity, quality and knowledge sharing. Regardless of all known benefits, achieving effective team collaboration remains a challenge. In some cases, substantial time and resources are consumed without yielding the desired benefits. Therefore, it is important to determine when collaboration is truly needed and in what intensity [1].

Software development is a complex process that involves the interaction of several people over a period of time to achieve a common goal. The team must communicate, share knowledge and artifacts, and coordinate the work [2, 3]. One challenge to collaboration support is the understanding of the interactions among developers.

We claim that collaboration in software processes can be turned explicit using social networks [4]. The exploration of existing networks among team members offers information that can help to understand and monitor the level of collaboration in a software project. As a consequence, it can improve the collaboration, communication and information flow.

For this purpose, a social network visualization and analysis tool – EvolTrack-SocialNetwork – was previously proposed. In this work, an exploratory study was planned and conducted, in the context of a free/open source software (FOSS) project, using EvolTrack-SocialNetwork to understand collaboration among team members. The preliminary results indicate a potential to enhance collaboration understanding.

V. Herskovic et al. (Eds.): CRIWG 2012, LNCS 7493, pp. 113–120, 2012.

The next section is dedicated to social networks. In Section 3, the main features of EvolTrack-SocialNetwork are presented. Section 4 describes the plan and the results of the exploratory study. Finally, Section 5 concludes the paper.

2   Social Networks

A social network (SN) is established when relationships with a defined semantic can be identified among a finite set of actors (nodes) [5]. The semantics depend on the analysis context (friendship, business, epidemics, etc.). Social network analysis (SNA) [6] enables to understand the relationships that facilitate or hinder the collaboration. It is possible to identify: individuals that carry out central roles; isolated groups or individuals; "bottlenecks" (indicate gatekeepers or brokers); etc. SNA represents a promising approach to collaboration understanding, being the main instrument used in this research work to diagnose collaboration in software teams.

2.1   Collaboration Characteristics in Social Networks

A vast range of measures of SN exists. In [4], centrality and density have been explored due to their potential to explain collaboration. *Centrality* is related to the intensity of involvement of a node in the network. *Density* concerns the network as a whole and measures the number of links that keep nodes interconnected [5].

Based on the interpretation and combination of these properties, Santos *et al.* [4] suggest three group *coordination characteristics*: in projects with *centralized* coordination, there are one or few big leaders who concentrate the activities. In *multiple* coordination, there are more central and intermediary nodes, which act like links among small subgroups within the network, sharing their tasks with other group members. Finally, in *distributed* coordination, collaboration is a constant and the existence of a leader is not a cornerstone. Group members balance their activities and there are no prominent actors in the network. Network density progressively increases over these characteristics.

These characteristics were defined by mapping different values of SN properties to the collaboration levels of the Collaboration Maturity Model (CollabMM) [7]. CollabMM describes an evolutionary path in which processes can progressively achieve higher capability on collaboration. CollabMM acts as a framework that defines the collaboration levels (Ad-hoc, Planned, Aware and Reflexive) and summarizes their main characteristics.

To illustrate these coordination characteristics, suppose a software project, which requires a high collaboration level. Considering CollabMM levels, this project would benefit if collaboration accomplishes a Reflexive level. The characteristic of distributed coordination represents this level. Therefore, a network similar to the one presented in Fig. 1a, which has a high density and relationships among nodes tend to be equally distributed, is expected. However, during project execution, the actual SN is looking like the one in Fig. 1b. In this centralized network, there is a strong leadership of one single node that is controlling tasks and information flow. Based on this information, it is possible to plan, using CollabMM collaboration levels and practices, how to achieve the desired collaboration outcomes for the project. This kind of information can be useful for project management and team knowledge.

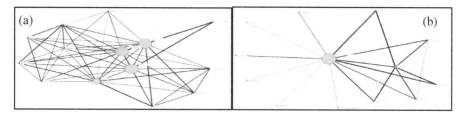

Fig. 1. Planned and Current and Social Networks

2.2   Features for Social Network Tools

There are several SN tools [8]. We analyzed 10 of them: UCINET, Pajek [9], Ariadne [10], Augur [11], MiSoN [12], OSSNetwork [13], RaisAware [14], Sargas [15], SVNNAT [16], and Visone [17]. Tools dedicated to software development analyses [18], but that do not specifically use SN, were not covered.

Considering these tools, and our research goals with SNA in software, a list of 15 desired features was gathered [8]. Some of these features include:

- Visualization of technical, socio-technical (ST) and social networks – contributes to understanding coordination and communication dependencies among developers that arise from technical dependencies in source code;
- Presentation of network evolution over time – contributes to the observation of network dynamic behavior and versions evolution.

The studied tools were analyzed in accordance with the complete features list and none of them completely satisfies these features. Some of the identified tools are not dedicated to the software development domain; do not engage in analysis specifically dedicated to collaboration; are not actually available or have significant limitations. Thus, there was room to propose another tool that supports collaboration analysis and brings new contributions. This motivated the creation of EvolTrack-SocialNetwork.

3   EvolTrack-SocialNetwork

EvolTrack-SocialNetwork[1] (Fig. 2) [19] purposes is to provide collaboration information that will be helpful to the understanding about the project collaboration level. For project manager, the tool provides useful information to track progress, conflicts, communication needs, etc. It opens the possibility of managing collaboration during project execution by making decisions and intervening in the ongoing work such as changing team members' roles or responsibilities. For developers, it offers awareness of other's work to provide context to their own activity. This improves the understanding of work dependencies.

---

[1] Site EvolTrack-SocialNetwork:
   http://reuse.cos.ufrj.br/evoltrack/socialnetwork

The *technical network* is extracted from mining source code from different CM repositories. In this network, all nodes are artifacts and edges represent a dependency relationship among them. It provides understanding of the software architecture.

From the dependencies and updates in the source code, it is also possible to know software developers who are responsible for the artifacts. When the technical network is annotated with this social information, it originates the *ST network* composed by actors and artifacts. This network integrates the coordination structure from the technical work unit (artifacts) with social information about authors and editors [3].

Because ST network describes both technical dependencies and authorship information, they can be used to generate SN describing the relationships only among software developers [10]. These relationships among developers exist because of dependencies in the source code they are working on and can have two semantics:

- **Dependency network:** represents social dependency among actors, due to the structural dependence among their artifacts;
- **Conflict network:** represents possible work conflicts among actors who have worked or are working on the same artifact.

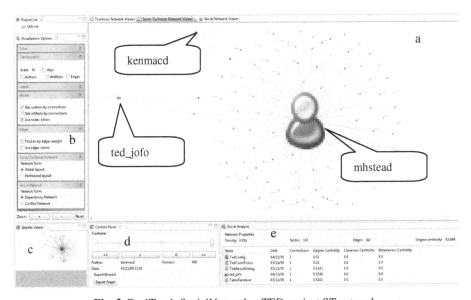

Fig. 2. EvolTrack-SocialNetwork – TED project ST network

In all these networks, different icons are used to highlight the changes or additions in the networks from one version to another. It supports the observation of network dynamic behavior and **versions evolution overtime**. In order to facilitate navigation among different network versions, there is a control panel (Fig. 2d).

Aiming to enhance the visualization options (Fig. 2b), one can apply: **filters** (name, number of connections, etc.); **a size scale** (increase nodes size proportionally to their number of relations); **transparency** (change the transparency level of nodes or edges that are recently acting in the network). Finally, the tool calculates nodes and network properties and displays them in the **analysis panel** (Fig. 2e). It allows identifying, for instance, the node that is a hub in the project.

4   Exploratory Study

This exploratory study has the following goal and scope:

Analyze collaboration in software development SN using EvolTrack-SocialNetwork tool
With the purpose of characterizing
With respect to effectiveness
From the point of view of collaboration researchers
In the context of free/open source software (FOSS) projects

A FOSS project was randomly chosen from the SourceForge portal, considering two prerequisites: Java as programming language; at least three developers to have a minimum of relationships. As result, the project Torrent Episode Downloader (TED) was selected. Data from TED project was extracted from the SVN repository. This project has more than 120 artifacts and 4 developers.

Five main instruments were designed and validated with an expert: the *term of consent* that declares the purpose of the study and ensures data confidentiality; the *characterization form* to determine the participants' profile; *training material* used to explain the main concepts of SN and describe the use of EvolTrack-SocialNetwork; the *study form* was used for collecting the results of the tasks; and a *survey*, which intended to obtain qualitative information about the study and the tool.

The study was conducted in 2011 with four participants, in individual sessions, over one week. Participants were 2 MSc students and 2 MSc in the collaboration area, of our research group, that did not have previous contact with EvolTrack-SocialNetwork. Their academic background was balanced and the participants' profile indicates a high level of experience with Software Engineering and Collaboration, but they are not experts in Project Management. In addition, participants stated, on average, low familiarity with FOSS projects or other SN tool.

During the study, participants used EvolTrack-SocialNetwork information to answer the 10 proposed questions. The tasks were divided into three groups with different levels of difficulty (Table 1). The most fundamental and simple tasks have been classified as *filter tasks*. *Basic tasks* are the ones that can be solved by extracting facts from the visualization or analysis. *Transfer tasks* required the participants to use their knowledge and reasoning in order to interpret the information. Each question was used to explore different perspectives on the visualization and analysis of SN. For instance, Q1 tested if participants were able to understand network participants.

Regarding tasks results (Table 1), most participants missed Q3, because there were classes with similar names and they focused on the first one they found (with the shortest distance). On average, participants correctly answered 8.25 questions in 22.75 minutes. Although it indicates a high level of effectiveness, it is only initial evidence, because of the limited number of participants in this study.

Although only one of the participants has incorrectly answered Q10, all of them faced difficulties to classify the network according to the coordination characteristics. In the used network all actors are connected, which characterizes a distributed coordination with highest density of the network. Therefore, we expected a better result in this last question. This result highlighted the complexity of these characteristics and gathered a new requirement: automatically classify the network.

Table 1. Tasks proposed to the study and their results

Id	Difficulty	Question	P1	P2	P3	P4
Q1	Filter	What actors are in version 1014 of the SN?	✓	✓	✓	✓
Q2	Filter	What is the density of version 1036 of the ST network?	✓	✓	✗	✗
Q3	Basic	What is the distance between TedMainDialog and TedMainToolBarButton in version 1036 of the technical network?	✗	✗	✗	✓
Q4	Basic	Who has the highest centrality in version 1036 of the ST network?	✗	✓	✓	✓
Q5	Basic	What has changed in the SN from version 900 to version 1014?	✓	✓	✓	✓
Q6	Basic	Who can mhstead conflict in version 900?	✓	✓	✓	✓
Q7	Transfer	What artifact has evidence that needs to be refactored?	✓	✓	✓	✓
Q8	Transfer	Who do you assign the work on the artifact GeneralPanel?	✓	✓	✓	✓
Q9	Transfer	With whom does ted_jofo need to coordinate the work?	✓	✓	✓	✓
Q10	Transfer	How would you define the coordination characteristic of the project?	✓	✗	✓	✓

At the end, participants provided qualitative feedback (Table 2) about the study. In general, they considered to be able to execute the tasks (Table 2a). Only P1 registered a difficulty with Q10. They were satisfied with the results obtained, but P3 faced difficulty to indicate refactoring without looking at the source code (Table 2b). All participants agreed that it is possible to understand how collaboration takes place in a project using the information presented. Participants' comments indicate difficulties (Table 2c) in defining the best type of network to be used in each kind of analysis.

Regarding the tool, participants rated it as easy or very easy to use (Table 2e). They indicated that EvolTrack-SocialNetwork greatly facilitated the proposed tasks (Table 2d). It was also mentioned that the visualization can be difficult in large networks, due to the overlapping of nodes and edges. Then, clustering and drill-down features are being considered for future versions.

Table 2. Participants' feedback

Item	P1	P2	P3	P4
a) Execution	Partially	Yes	Yes	Yes
b) Satisfaction	Yes	Yes	Partially	Yes
c) Awareness	Yes	Yes	Yes	Yes
d) Contribution	+ facilitated	+ facilitated	+ facilitated	+ facilitated
e) Difficulty	Easy	Very easy	Easy	Easy

Regarding *threats to validity,* we should notice that the size of population is reduced, which prevents the use of quantitative analysis and results generalization. In addition, subjects are members of the same research group of the authors. Participants had the expected profile for this study, but their low experience with project management can influence the decisions in tasks that they were not used to perform.

The grouping of tasks by type helps to analyze the data, but the same weight was assigned to tasks with different levels of difficulty. Finally, the study considers only a single FOSS project. Thus, experiments with a larger quantity or other types of projects should be executed. Another opportunity is to show the networks to the chosen FOSS community and get them to reflect and comment on its accuracy. It can complement the results, obtained through data mining and networks generation, with perspectives acquired via observation or interviews to provide a deeper interpretation.

5   Conclusion

This paper presented an initial exploratory study for understanding the collaboration in software development, through SN using EvolTrack-SocialNetwork as a support tool. Results indicate that participants show a positive feedback about the tool potential to increase their ability to be aware of collaboration in a software project. Refinement of the collaboration levels/characteristics were also identified as needed in order to improve the tool effectiveness. The study also helped to identify further requirements to improve the tool usability.

The exploratory study presented in this paper is part of a broader evaluation plan. After implementing some of the ideas gathered from this study, our research agenda comprises a further feasibility study with 10 participants (project managers coming from industry) and using two FOSS projects. The purpose is to assess the EvolTrack-SocialNetwork support on project managers' decision making. The plan includes comparing the use of EvolTrack-SocialNetwork with other typical Software Engineering artifacts, such as a class diagram and a configuration management log.

Acknowledgments. This work is partially funded by CNPq (n°. 142006/2008-4 and 310776/2009-0) and is part of INCT Program, supported by CNPq (n°. 557.128/2009-9) and FAPERJ (n°. E-26/170028/2008). The authors would like to thank Marcelo Schots for the revision of the study plan and participants for their time and contributions.

References

1. Hansen, M.T.: When Internal Collaboration Is Bad for Your Company. Harvard Business Review 84, 83–88 (2009)
2. Mistrik, I., Grundy, J., Hoek, A., et al.: Collaborative Software Engineering. Springer, Heidelberg (2010)
3. Valetto, G., Helander, M., Ehrlich, K., et al.: Using Software Repositories to Investigate Socio-technical Congruence in Development Projects. In: International Workshop on Mining Software Repositories (MSR), Minneapolis, USA, pp. 25–28 (2007)
4. Santos, T.A.L., Araujo, R.M., Magdaleno, A.M.: Bringing Out Collaboration in Software Development Social Networks. In: International Conference on Product Focused Software Development and Process Improvement (PROFES) - Short Papers, pp. 18–21. ACM, Torre Canne (2011)

5. Wasserman, S., Faust, K.: Social Network Analysis: Methods and Applications. Cambridge University Press, Cambridge (1994)
6. Barabasi, A.L.: Linked: How Everything Is Connected to Everything Else and What It Means for Business, Science, and Everyday Life. Plume, Cambridge (2003)
7. Magdaleno, A.M., Araujo, R.M., de Borges, M.R.S.: A Maturity Model to Promote Collaboration in Business Processes. International Journal of Business Process Integration and Management (IJBPIM) 4, 111–123 (2009)
8. Magdaleno, A.M., Werner, C.M.L., de Araujo, R.M.: Analyzing Collaboration in Software Development Processes through Social Networks. In: Margaria, T., Steffen, B. (eds.) ISoLA 2010, Part I. LNCS, vol. 6415, pp. 435–446. Springer, Heidelberg (2010)
9. de Nooy, W., Mrvar, A., Batagelj, V.: Exploratory Social Network Analysis with Pajek. Cambridge University Press, Cambridge (2005)
10. de Souza, C.R., Quirk, S., Trainer, E., et al.: Supporting collaborative software development through the visualization of socio-technical dependencies. In: ACM SIGGROUP Conference on Supporting Group Work, pp. 147–156. ACM, Sanibel Island (2007)
11. de Souza, C., Froehlich, J., Dourish, P.: Seeking the source: software source code as a social and technical artifact. In: International ACM SIGGROUP Conference on Supporting Group Work, pp. 197–206. ACM, Sanibel Island (2005)
12. Aalst, W., Reijers, H.A., Song, M.: Discovering Social Networks from Event Logs. In: Computer Supported Cooperative Work (CSCW), vol. 14, pp. 549–593 (2005)
13. Balieiro, M.A., Júnior, S.F.S., Souza, C.R.B.: Facilitating Social Network Studies of FLOSS using the OSSNetwork Environment. In: Open Source Development, Communities and Quality, pp. 343–350. Springer, Boston (2008)
14. Costa, J., Feitosa, R., de Souza, C.: Tool support for collaborative software development based on dependency analysis. In: International Conference on Collaborative Computing: Networking, Applications and Worksharing, pp. 1–10. IEEE, Chicago (2010)
15. de Sousa, S.F., Balieiro, M.A., dos R. Costa, J.M., et al.: Multiple Social Networks Analysis of FLOSS Projects using Sargas. In: 42nd Hawaii International Conference on System Sciences (HICSS), pp. 1–10. IEEE (2009)
16. Schwind, M., Schenk, A., Schneider, M.: A Tool for the Analysis of Social Networks in Collaborative Software Development. In: Hawaii International Conference on System Sciences (HICSS), Koloa, Kauai, Hawaii, United States, pp. 1–10 (2010)
17. Brandes, U., Wagner, D.: Visone - Analysis and visualization of social networks. In: Graph Drawing Software, pp. 321–340. Springer, Heidelberg (2003)
18. Ogawa, M., Ma, K.-L.: Software evolution storylines. In: 5th International Symposium on Software visualization, pp. 35–42. ACM, New York (2010)
19. Vahia, C.M., Magdaleno, A.M., Werner, C.M.L.: EvolTrack-SocialNetwork: Uma ferramenta de apoio à visualização de redes sociais. In: Congresso Brasileiro de Software: Teoria e Prática (CBSoft) – Sessão de Ferramentas, São Paulo, SP, Brasil, pp. 7–13 (2011) (in Portuguese)

Keeping Up with Friends' Updates on Facebook

Shi Shi, Thomas Largillier, and Julita Vassileva

MADMUC Lab
University of Saskatchewan
Saskatoon, Canada
shey.shi@usask.ca,
{thomas.largillier,julita.vassileva}@usask.ca

Abstract. Users of social network sites, such as Facebook, are becoming increasingly overwhelmed by the growing number of updates generated by their friends. It is very easy to miss potentially interesting updates, it is hard to get a sense of which friends are active and especially, which are passive or completely gone. Such awareness is important to build trusted social networks. However, the current social network sites provide very awareness of these two kinds.

This paper proposes a interactive method to visualize the activity level of friends. It creates a time- and an activity-pattern awareness for the user, as well as an awareness of the lurkers. The proposed visualization help the user to browse her friends depending on how recently they have posted and how much interactions their updates have caused.

1 Introduction

Social Network Sites (SNS) have experienced an explosive growth in recent years. There are more than 845 million active users on Facebook, and more than 57% of them log on to Facebook on any given day[1]. A large proportion of these users share updates of their status with friends, including messages about their thoughts, their current location, links to interesting articles or videos, statements of activities (e.g. they have befriended other users, or the messages generated as a side effect of playing games). Such status updates will be called "social data" in this paper. A large amount of social data is generated every day, which triggers an information overload for users. However, it is very easy to miss something important or interesting, if one has not logged in for one or two days, and a lot of updates were shared by her friends during this period. Also it is not easy to find out if a particular user has posted something recently, or who is generally active on Facebook and who is just a lurker. Therefore, it is necessary to provide a better way to organize and present social data to make users aware of the pattern of online social activities. Information visualization technology can provide effective approaches for presenting large amounts of data intuitively, which can help users to get insights into the data, discover patterns and find information of interest easily.

---

[1] http://newsroom.fb.com/content/default.aspx?NewsAreaId=22

V. Herskovic et al. (Eds.): CRIWG 2012, LNCS 7493, pp. 121–128, 2012.

This paper proposes an intuitive and easy to understand visualization method for data streams that creates the needed awareness for the user about her social network. The implemented visualization application was designed for Facebook and provides navigational and interactive methods to access posts of all the user's friends, so she can browse the social data shared by her friends more easily and selectively.

The rest of the paper is organized as follows: section 2 presents an overview of existing social visualization tools. Section 3 describes the conceptual design and implementation of the proposed approach. Section 4 presents the future work and concludes the paper.

2   Related Work

Social visualization can be defined as the visualization of social data for social purposes [4]. In other words, social visualization uses information technology and focuses on people, groups and their conversational patterns, interactions and relationships with each other and with their community [4]. Visualizations of social data can be used for increasing awareness of one's social activities, motivating users to participate in social communities, and coordination. There are various social visualization approaches and techniques that have been proposed.

The Babble system [2] is one of the first approaches integrating the social visualization technology into an online chat room system. Each person in the system is represented by a dot of different color. A gray circle in the center of the visualization represents the proxy of the current chat room. All users, who have already logged in to the system, but not in the current chat room, will be positioned outside the gray circle. The dots located inside the circle denote users who are in the current room. When people are active in the conversation their dots move to the center of the circle, and then drift back out to the edge when they stop talking for 20 minutes.

Comtella [6] is a file-sharing community that uses a metaphor of a night sky in which every user is represented by a star. The size of the star indicates a user's number of contributions. A star with more red hue (warmer) represents a user who has shared more new files than the number of downloaded files from other users, and a star with more blueish hue (colder) represents a user who downloaded more files than she has shared in the community. The big yellow star represents the "best user" who shares more than everyone else and has contributed new things to a community. Therefore, the visualization encourages social comparison among users to increase the diversity of resources in a community.

Data portraits are very useful for this purpose. For example, PeopleGarden [8] is designed for online interaction environments such as web-based message boards, chat rooms, etc. In PeopleGarden, a flower metaphor, including magenta petal (for initial post) and blue petal (for response) has been used for each user in the system. Dots on the petal indicate the number of answers to this post. The height of the flower reflects how long the user has been in the system.

Faded petals are used to indicate old posts. People may be motivated by the visualization to post more, and in this way get more petals for their flower.

IBlogVis [3] uses the digital footprints method to help a user find interesting articles when she is browsing blog archives. In IBlogVis, each blog entry is displayed as a point on the time line located in the middle of the page. A vertical line above each point represents the length of each entry, and a second vertical line below each point represents the total length of comments this entry has collected. The circle's radius on the end of this line indicates the number of comments for each entry. This visualization application provides a rich overview of a blog.

Data streams visualization mainly focuses on high throughput streams and the objective is to visualize trends in the stream. Wong *et al* in [7] present two methods that can be combined to visualize data streams. Their methods are based on multidimensional scaling and represent objects from the stream as vectors. The vectors are then displayed on a plane. When dealing with text streams, their method can be used to extract the topics discussed in the stream. This method is designed to display a large quantity of data on a screen and therefore cannot embed all the information required by the human reader.

Facebook, as one of the most popular online communities, also has some visualization applications to help its users to explore their social data. All visualizations for Facebook, like Facebook Social Graph[2], Facebook Friend Wheel[3], Facebook visualizer[4] or Nexus[5] offer the user a better representation of her social network by organizing her friends and their relations or affinities in a graphical way. To the best of our knowledge, there is still no social data streams visualizations that allow users to efficiently browse their social data.

All those visualization methods may help the user have a better understanding of her social network or the activity occurring inside a community but none help her browse a data stream in a more efficient way. Users spend a lot of time browsing over updates they don't care about or that they might already have read if those which are popular stay on top of the representation.

This paper proposes an interactive visualization approach for Facebook's social data stream that allows discovering the time patterns and the main current contributors, as well as the lurkers.

3   Proposed Visualization

Classic data streams present social data in a reverse-chronological order. In order to ease its users navigation and not force them to browse the entire stream before finding something interesting, Facebook reorganizes the social data using its "top stories"[6] mechanism. Even without considering the fact that the "top stories"

---

[2] http://www.mihswat.com/labs/app/facebook-social-graph
[3] T. Fletcher. Friend Wheel. http://thomas-fletcher.com/friendwheel/, 2009.
[4] http://vansande.org/facebook/visualiser/
[5] http://nexus.ludios.net/
[6] http://www.facebook.com/help/?faq=277741542238350

might not be accurate for not regular Facebook users, or users that didn't fully fill their profile, the social data is still presented in a stream and includes all its limitations. First it is impossible to get an overall picture of who has posted updates recently, how recently, how many updates, and which of the friends have not been active, since only a couple of stories are presented on the screen at any given time. Also it may be overwhelming to view the posts if the user has not logged in for a long time, or if her friends have been very active during her offline period. It is also very easy to miss posts that could be potentially interesting either because they are too low in the stream or not have been selected as important by the mechanism. While Facebook provides the option to check the updates of a specific friend, it is not easy, since only a few friends are presented at a time on the screen, and to find a particular friend, a user has to search for him/her.

For these reasons, it is important to make users visually aware of who has posted and at what time, the number of posts, i.e. how active the user has been recently and how popular the posts are, i.e. how many likes and comments were received by each post.

The goal of the proposed visualization, called Rings[7], is to ensure an alternative way of browsing the stream of social data on Facebook, which allows the user to see which of her friends has been active recently, who posted many popular updates recently or not and who has stopped sending updates. This will reduce the cognitive overload of the user and will allow her to quickly check posts by particular friends, to be aware of (and possibly ignore) the most active users, and also to be aware of the users who are not posting and may be lurkers.

The design includes each individual user's representation in the visualization (for simplicity, it will be called "avatar"), visualization layout, functions, and application user interface. The avatar focuses on how to reflect the number of posts from a user during last 30 days in the visualization. How to arrange a large quantity of avatars in a neat and appealing way is a challenge that the visualization layout has to address. Rings' user interface and functions aim at providing an easy way for the user to navigate in the visualization and access the usual Facebook content through it.

Avatar Visualization. In Rings, each user is represented as a spiral. The number of the posts in last 30 days is scaled into one of the six different levels of contribution. To visualize these levels, different sizes and colors of spirals are applied to represent the six levels (see Fig. 1).

In addition to this, the related usability research shows that approximately 10% of human males, along with a rare sprinkling of females, have some forms of color blindness[8]. Thus, the six colors are carefully chosen and tested under all the forms of color blindness on Colblindor[9]. In order to help users recognize their friends more easily, the profile picture and the username of each user on

---

[7] http://rings.usask.ca

[8] A. Wade. Can you tell red from green? http://www.vischeck.com/info/wade.php, 2000

[9] http://www.colblindor.com

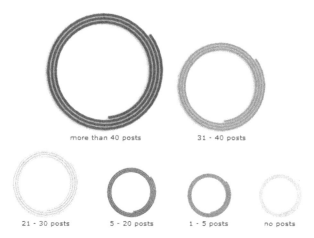

Fig. 1. Each level of quantity is indicated with a specific color (red, orange, yellow, green, blue, gray) and size

Fig. 2. The profile picture and the username are displayed in the spiral

Facebook are displayed in the spiral, along with the number of posts the user has contributed during the last 30 days as shown on Fig 2.

In order to see the posts made by one of her friends, the user only has to hover the mouse over her friend's avatar to see a detailled list of her friend last 30 days activity. To reflect how interesting/popular posts are, the numbers of likes and comments they receive are used. According to the total number of likes and comments, each post is classified into 5 different popularity levels displayed with different emphasis on the screen by means of different shades of gray. All the 5 levels are presented with 5 different gray colors [1]. For example, a post with many likes and comments is shown in solid black color, while a post with no likes or comments is shown in light-gray color. Additionally, to indicate the exact numbers of likes and comments, a bracket with two numbers is added at the very beginning of each post in the floating window if there are some likes and comments for this post. For instance, [L:4 C:3] means there are 4 likes and 3 comments on this post.

This strategy is also applied to the avatar visualization on the screen to provide awareness for the user to see at a glance which Facebook friends have some interesting/popular posts. As discussed in the last paragraph, each post is classified into one of the five different popularity levels according to the total number

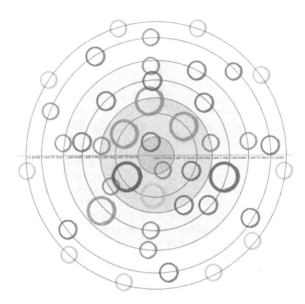

Fig. 3. Layout of the visualization

of likes and comments. Similarly, the avatar visualization is also classified into one of the five different popularity levels according to the highest popularity level of posts that the user has got and five different opacity levels are used to present the five different popularity levels of users.

Layout. The number of friends varies drastically among Facebook users. For example, there are quite a few users with over 1000 friends. Considering the acceptable loading time, the unavoidable timeouts of the Facebook API, and the resulting crowded screen, it is impossible to display all the friends of such a user on the screen at the same time. Therefore a restriction was introduced in the design on the number of friends that can be displayed in one screen. If a user has more than 200 friends on Facebook, they will be separated in groups of equal size $k < 200$. The user can select any of these groups to display. Then the visualization will only display these selected k friends after an acceptable loading time. The groups are balanced to prevent one screen to be packed while one has only a few users.

In order to represent how much time has elapsed since the latest post by a specific friend, the background layout was designed as a set of concentric rings, where the friends who have posted most recently are displayed in the center, and people who have posted long time ago will be shown at the periphery. There are several rings on the screen to indicate different time periods in the past. The rings, from the center to the periphery, show the last 3 hours, last 12 hours, last 24 hours, last 3 days, last week, last 30 days, and no posts. We chose to display the activity of the user's friends during the last month, since it is a long enough period to realize who among her friends has stopped posting updates.

Indeed, some users might stop posting during a couple of weeks because they are on vacations, while no activity during a whole month is much more significant. Also the next scale was to display a whole year of activity which is way too much for any user to browse efficiently her friends' social data. Each avatar representing a Facebook friend is placed on a specific ring according to the post-time of her latest post as represented on Fig 3. For example, a user will show up at the very center in the visualization if she posted something in the last 3 hours. If she stops posting anything from then on, her avatar will keep drifting to the periphery in the visualization over the next 30 days and will finally settle somewhere on the outmost ring.

Since research shows that humans naturally tend to focus their attention to the center of an image, the user's attention will focus on the most recently active users, similar to the default display option in most streams (the most recent or most popular at the top). This design also naturally focuses the attention of the user to the center (the "Bull's Eye"), where the action is, and the most recent posts are.

The concentric rings design allows for scalability, since the time periods represented with concentric circles are getting longer as they are getting further from the center of the visualization. There will be fewer people who posted very recently and the space in the center is limited, while there will be many more people who have posted in the past, the more distant the past, the larger the ring and more space available to accommodate more avatars without being crowded.

4   Future Work and Conclusion

This paper introduced an intuitive and interactive visualization creating an increased awareness in the user about her social network on Facebook and allowing her to get insight about the level and pattern of posting activities of her friends. It provides an alternative way to browse Facebook's social data stream.

The next step is to conduct a large-scale evaluation of Rings. The objective of this study will be to show that the Rings visualization helps users access the information they are interested in faster and are not bother by the noise in their social data stream. The authors have already conducted small scale studies that showed promising, while not statistically significant, results.

We also need to improve some elements of the current of Rings. Currently if a user possess more than 200 friends, they will be automatically separated into balanced groups of $k < 200$ users. This separation is currently done without any specific algorithm to regroup cluster of users together and cannot be influenced by the user. The first modification is to authorize users to rearrange groups as they like while developing an algorithm that will regroup more connected people together.

Also there are several visual parameters currently unused by Rings. First it will be really interesting to explore the angle in the visualization to position the avatars in proximity to each other, depending on different criteria, e.g. if they are friends with each other (in this way, it will create awareness of the structure of social network), or if they belong to the same organization, or share similar interests (addressed by other social graph visualizations, e.g. for LinkedIn). Various criteria for proximity can be used. In order to keep the main focus of the visualization on the time pattern of posts, the proximity would be secondary to the time pattern of posts, which is the main criterion for arranging the avatars on each ring.

Rings could take into account evidence of other user activities, such as liking or commenting, or just logging in or scrolling, rather than just number and recency of updates. This would require enhancing the visual language to distinguish visually the different forms of activity. It would be an important extension since many online community users don't consider themselves lurkers, if they read, comment or rate [5].

Finally, applying a similar visualization to other social network sites, such as Google+, Twitter or LinkedIn and creating an aggregator for all the users' social network sites is a natural extension of this work.

References

1. Alexander, J., Cockburn, A., Fitchett, S., Gutwin, C., Greenberg, S.: Revisiting read wear: analysis, design, and evaluation of a footprints scrollbar. In: Proceedings of the 27th International Conference on Human Factors in Computing Systems, pp. 1665–1674. ACM (2009)
2. Erickson, T., Kellogg, W.A.: Social translucence: an approach to designing systems that support social processes. ACM Transactions on Computer-Human Interaction (TOCHI) 7(1), 59–83 (2000)
3. Indratmo, J., Vassileva, J., Gutwin, C.: Exploring blog archives with interactive visualization. In: Proceedings of the Working Conference on Advanced Visual Interfaces, pp. 39–46. ACM (2008)
4. Karahalios, K., Viégas, F.: Social visualization: exploring text, audio, and video interaction. In: CHI 2006 Extended Abstracts on Human Factors in Computing Systems, pp. 1667–1670. ACM (2006)
5. Nonnecke, B., Preece, J.: Lurker demographics: Counting the silent. In: Proceedings of the SIGCHI Conference on Human Factors in Computing Systems, pp. 73–80. ACM (2000)
6. Sun, L., Vassileva, J.: Social visualization encouraging participation in online communities. In: Groupware: Design, Implementation, and Use, pp. 349–363 (2006)
7. Wong, P.C., Foote, H., Adams, D., Cowley, W., Thomas, J.: Dynamic visualization of transient data streams. In: IEEE Symposium on Information Visualization, INFOVIS 2003, pp. 97–104. IEEE (2003)
8. Xiong, R., Donath, J.: Peoplegarden: creating data portraits for users. In: Proceedings of the 12th Annual ACM Symposium on User Interface Software and Technology, pp. 37–44. ACM (1999)

Formal Modeling of Multi-user Interfaces
in Cooperative Work

Benjamin Weyers[1], Wolfram Luther[1], Nelson Baloian[2], Jose A. Pino[2]

[1] University of Duisburg-Essen, Department of Computer Science and Cognitive Science,
Lotharstr. 63, 47057 Duisburg, Germany
[2] Universidad de Chile, Department of Computer Science,
Blanco Encalada 2120, Santiago, Chile

Abstract. Support systems for cooperative work lack consistent modeling tools for user interface creation and execution that are flexible enough to combine both data processing and the logical aspects of a user interface and, at the same time, dialog and cooperation modeling aspects. This paper introduces a new concept to model user interfaces for cooperative work: the so-called multi-user interfaces aimed at distributed scenarios involving mobile devices implementing cooperative work. These multi-user interfaces are modeled in a hierarchical structure of dialog models and interaction logic based on a formal modeling language called FILL. For execution and verification, FILL models are automatically transformed to reference nets, a type of Petri nets, making the entire user interface and cooperation model accessible to simulation and verification tools. This new approach seeks to integrate more closely modeling and implementation based on a formalized interface design and user-machine dialogue. Formal graph rewriting concepts allow both the user interface and the collaboration model to be easily adapted in various ways by the modeler or user.

Keywords: Multi-user interfaces, mobile cooperative work, formal UI models.

1   Introduction

The research field of computer-supported cooperative work (CSCW) is mainly concentrated on examining "the possibilities and effects of technological support for humans involved in collaborative group communication and work processes" [4, p. V]. Key research areas in CSCW include the development of concepts and architectures for communication, cooperative interaction concepts, and ubiquitous computing approaches for hiding technology from the user and promoting integration of cooperative work in everyday working processes.

Up to now, there is no all-embracing formal concept for modeling user interfaces (UI) in CSCW scenarios for creating flexible UI models. Formal UI models can be directly processed by algorithmic implementations, also involving further formal models, e.g., user models, dialogue or context models, making this kind of UI implementation suitable for machine processing. The creation of flexible UI models would make CSCW systems more adaptable and easy to integrate into existing workflows

V. Herskovic et al. (Eds.): CRIWG 2012, LNCS 7493, pp. 129–136, 2012.

than current systems. It would also provide a formal basis for validation and verification concerns, thus identifying problems in cooperation. Finally, it would enable a close and error-free combination of modeling and implementation. To this end, we are introducing an extension to our formal modeling approach for UIs called FILL [15] that responds to the five basic problems in CSCW design mentioned by Koch and Gross in [9]: awareness, communication, coordination, consensus, and collaboration. We do not claim that our general modular architecture for formal (multi-)UI modeling based on FILL offers final UI patterns or concrete UI solutions for CSCW scenarios. Nevertheless, this paper presents a conceptual basis for future implementation by indicating the capabilities of a formal approach to model UIs for CSCW systems, integrating CSCW mechanisms into a UI model, and exemplarily demonstrating its potential use.

FILL-based UI models are mainly based on interaction logic (IL), a term that describes the data processing between, on the one hand, the participating users carrying out actions on the UI physical representation (PR) (e.g., buttons, sliders) and, on the other hand, the system being controlled in the form of system information data sent by the system to the user. Thus, the new approach introduced in this contribution fills the gap between classic informal or semi-formal UI modeling and its implementation by using a visual modeling language accompanied by an automatic transformation to an executable and verifiable formal model, Petri nets. Formal reconfiguration concepts make adaptation of CSCW systems more comprehensive when modifying an existing model at a later stage. Besides addressing the concerns of CSCW systems and groupware modelers, this approach enables the user to introduce CSCW principles into the formal reconfiguration concepts that specifically affect the cooperation process.

This paper will begin with a short insight into related research fields, followed by an introduction to FILL in Section 3. Section 4 and 5introduce the extension of the basic UI modeling approach to cooperative work scenarios and describe briefly an example application supporting cooperative learning of cryptographic protocols.

2   Related Work

User interface modeling plays a central role in the development of groupware and CSCW systems, especially in the context of implementing certain CSCW features, like awareness, communication, etc. Various researchers suggest ways of creating and modeling user interfaces for groupware. Molina et al. [11] present, e.g., a model-based methodology for the creation of user interfaces for CSCW systems by combining high-level specification concepts defining cooperative tasks with a model of the organization. These models are iteratively refined, resulting in ConcurTaskTrees serving as bridges between the conceptual model of the CSCW system and its user interface implementation. Arvola [1] introduces a collection of interaction design patterns for CSCW systems that are based on field studies in face-to-face scenarios. These concepts can be used in ongoing modeling approaches as discussed in this contribution. Johannsen [8] describes general aspects in human-machine interface modeling for cooperative work in various use cases, such as cooperation in a cement plant.

A further key issue in CSCW systems modeling is how to bring CSCW principles into an existing formal user interface modeling approach based on Petri nets. Petri Nets have been already used to describe how CSCW systems work, especially from the user-machine interaction point of view [12]. Hao et al. [7] offer an example of Petri net–based modeling of heterogeneous data sharing in collaborative work: they describe the use of Petri net translation and firing rules in relation to a heterogeneous data queue as an access station to heterogeneous data sharing. Borges et al. [3] used Petri net–based models to describe communication in learning scenarios to be executed in an e-learning environment called TeleMeios. Furuta et al. [6] introduce a modeling approach based on colored Petri nets for modeling protocols for collaboration. Nielsen et al. [2] give a general overview of the use of Petri nets in modeling interactive systems. This list motivates our work, in which we seek to combine Petri nets with formal user interface modeling as introduced in [15].

3　Formal Modeling of Single-User Interfaces

In this section, the formal modeling approach for user interfaces used in a single-user scenario is highlighted, while necessary extensions to this basic approach will be outlined in Section 4, based on the entire formalization developed in detail in [15, 16]. Thus, the present section does not consider any aspects of CSCW systems to be introduced in the user interface model, since these will be discussed in Section 4.

A two-layered architecture has been developed for formal modeling standard single-user interfaces, as can be seen in Figure 1. It distinguishes a physical representation of a UI as a set of interaction elements (like buttons, sliders, text fields, etc.) from its interaction logic. Interaction logic is a set of data processing routines describing the user interface behavior. Thus, interaction logic describes what happens if the user presses a button or evokes another type of interaction event, as well as necessary data processing for representing information emitted by the (technical) system to be controlled. Interaction logic also models the interrelation among the various interaction elements of the physical representation involving data emitted from the system. This interrelation is normally identified as the user interface dialog model. In this modeling concept, the system itself is handled as a black box with a specific set of values to be read or written from the external interaction logic. This well-defined, finite set of system values is referred to as the system interface.

In [15], a graph-based language for modeling formal interaction logic called FILL was introduced. FILL is based on concepts familiar to business process modeling languages, which is also visible in certain nodes borrowed from BPMN, especially gateway nodes for fusion and splitting of processes. FILL has three node types: (a) operation nodes, (b) proxy nodes, and (c) the BPMN nodes mentioned above. Type (a) nodes represent values to be read by or written into the system (system operation nodes), including elementary data processing functions like type conversions, arithmetic operations, etc. (interaction-logic operation) and channels allowing the modularization of IL. Type (b) nodes represent

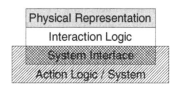

Fig. 1. Simple architecture of a formal user interface model

interaction elements in the UI's physical representation in the IL graph. Thus, proxy nodes handle data objects resulting from interaction, such as events and data returned from the interaction logic to certain interaction elements. Type (c) nodes are used for branching and fusing data processes in interaction logic models offering various semantics and guard conditions, which control the branching and fusing of data processes.

A formal algorithmic transformation of FILL models into reference nets has been implemented [15, 16] to offer formal semantics, executability, and a broad toolset for simulation and verification in FILL-based IL models. Reference nets are a special type of colored Petri net [10] with tokens representing complex data types; it is extended by an annotation language for controlling the switching of transitions based on logical expressions and tokens which are able to reference other net instances or different formal models, thus allowing modular FILL models. In general, reference nets can represent the entire semantics of FILL models, such as time discreteness, parallelism, and non-determinism. In conclusion, a (single) formal FILL-based UI can be defined as follows:

Definition 1: A user interface *UI* is a three tuple *(PR; IL; SI)*, where *PR* is the physical representation, *IL* is a FILL-based interaction logic or its representation as a reference net, and *SI* is a given system interface.

Adaptation of CSCW systems and their UIs is well supported by the use of formal languages for modeling UIs. Through the integration of CSCW models into formal multi-UI models (as described below), formal rewriting can be used to adapt interaction logic and its corresponding CSCW model. Graph-rewriting systems have been shown to be well suited to formally changing graph-based models [16] as in the double pushout approach introduced by Ehrig et al. [5].

For creation, simulation, and reconfiguration of UIs, a framework was implemented called the UIEditor [15]. It offers interactive editors for creating physical representations and FILL-based interaction logic, a simulation engine that combines Renew (http://www.renew.de) for the simulation of reference nets and the physical representation of user interfaces with a module for interactive or automatic reconfiguration based on pre-implemented algorithms. The creation module includes an algorithm that automatically translates a FILL model into its reference net–based representation. Examples can be found in [16] on pp. 43-45, pp. 86-87, and pp. 117-124.

4   Multi-user Interface Modeling

Based on the findings reported in [14] about too abstract architectures for modeling UIs, it is interesting to extend the proposed basic architecture to one offering finer grained exchangeable components and identification of dialog model components and their possible representation as FILL models. The resulting component-based modeling approach is necessary for the creation of multi-user interfaces, which are mainly characterized by a multi-user dialog model describing the interaction among multiple UIs on a higher meta level independent of the individual users' UIs. This meta model will contain formal (Petri net–based) descriptions of CSCW concepts, such as awareness, and make them executable, verifiable, and adaptable through reconfiguration.

A first step towards component-based modeling of user interfaces has been made using FILL by defining sub-processes called *interaction processes*, which are sub-graphs of the FILL-based IL that are exclusively associated with one interaction element of the physical representation. The interrelation among different interaction elements is part of the interaction processes inasmuch as they are connected by channels to each other. This modeling approach produces fixed dialog models entangled in the resulting IL. The concept of interaction processes has to be conceptually extended through a dialog model that describes the interrelation between the interaction processes using FILL. This will be realized by presenting interaction processes in the independent, FILL-based, higher-level dialog model with interaction-logic operation nodes, where input and output ports represent input and output channel operations defined in the interaction process. This modularizes the IL into interaction processes (one for each interaction element), and the dialog model makes them exchangeable. This FILL-based (local) dialog model is also transformed into a reference net, where the communication between dialog model and interaction process is described by the intrinsic ability of reference nets to reference other nets through synchronous channels [10].

Multi-user interfaces are user interfaces that are used simultaneously by more than one user. Various scenarios seem to be possible. The most common is probably the one-interface–multi-user scenario, in which one physical user interface exists, but several users use it, e.g., UI design in case of the control room of power plants, which can also be handled with the originally published FILL-based modeling approach. Another possibility is the multi-interface–multi-user scenario. Here, cooperative systems implemented on mobile devices are of special interest. In such systems, every mobile device has its own user interface used by one participant in the cooperative working scenario. Due to the capability of modularizing UI models with FILL, as will be introduced below, the example scenario of cooperation through mobile devices is an excellent use case demonstrating the possible distribution of device specific UIs using a formal modeling approach.

In Figure 2, the major architectural concept can be seen, differentiating n different mobile devices with each having a formally modeled user interface with a physical representation and a modularized interaction logic. A further abstraction layer of the dialog model has been added describing the communication among various participants and their mobile devices, called *multi-user dialog* or *cooperation model*. This global dialog model again is shown as a FILL-based model describing the data flow that is implemented in the underlying communication structure, such as a client-server or some other communication structure, for instance, Blackboard or peer-to-peer concepts. At the level of local and global FILL models, communication is based on the reference mechanisms of the reference nets resulting from the algorithmic transformation of FILL.

For modeling this kind of cooperation-handling model using FILL, every data type that can be sent to and from a client is indicated as a proxy node in the FILL-based,

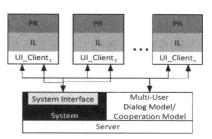

Fig. 2. Multi-user interface model

multi-user dialog model and controlled by the local interaction logic model of their local UIs. To distinguish the various participating clients and to identify certain rules in the dialog model, the sent data should be accompanied by a client ID. Since FILL is able to handle complex object-oriented data types, it is also possible to use role identifiers to get around the problem of static ID definition, which has to be integrated into the dialog model, but this makes it inflexible.

Processed data to be returned to the client can also be tagged with various kinds of information, for instance, whether a piece of data is awareness information. In general, this approach makes the model independent of the underlying technical communication structure. To avoid any inflexibility in the model, the graph rewriting technique mentioned above can be applied to the finished dialog model. This is the case if changes have to be carried out on the cooperation concept, like changing privacy rules or adapting aspects of awareness. Furthermore, client-side local IL models can also be modified using rewriting that can be triggered by the multi-user dialog model or through other instances, like supervisors or agents. Finally, multi-user interfaces can be defined as follows:

Definition 2 (Multi-User Interface): A multi-user interface MUI is a triple (U^*, MD, SI), where U^* is a set of modularized single UI models, MD is a FILL-based multi-user dialog model, and SI is a system interface. Every u_i in U^* has the form $u_i=(IP, LD)$, where IP is a set of interaction processes and LD is the local dialog model.

5   Modeling CSCW Systems Based on MUIs

Koch et al. [9] consider awareness, communication, coordination, consensus (here split up in authorization/privacy, and coordination/delegation as tools for finding consensus in cooperative work, cf. [9]), and collaboration as the five central issues to be tackled while developing CSCW systems.

Communication, inasmuch as it is based on data exchange, can be modeled using FILL. Here, the sent communication data can be filtered, redirected to different cooperation partner, or modified if necessary. To this end, specific interaction-logical operations defining how data is spread to other cooperation partners and how it is modified can be introduced. This procedure can be used to control the use of shared workspaces. It can also be used to manage data on which participant is able to recognize which kind of changes are being made in the workspace.

Awareness information, such as editing events in shared data, is processed by the underlying system holding the data, and by the multi-user dialog model, which can handle the creation and distribution of certain awareness information and send these to authorized cooperation partners. Furthermore, it is possible to change a participant's UI to that of a cooperating user, permitting access to that user's work and vice versa. This temporary sharing of UIs would also be helpful for providing information concerning tasks performed by group members and reflected in an operation accessible on a foreign UI.

Authorization and **privacy** can be defined in the multi-user dialog model by including specific interaction-logic operations that filter awareness or communication data by applying predefined rules. These rules might be defined by a supervising participant who is also able to control and supervise the entire cooperative work in the system. Reconfiguration can be used to adapt the privacy model during runtime, by, for instance, changing filter rules or various clients' UI models to enhance or restrict rights in altering data or perceiving changes in the system.

Coordination and **delegation** can also be modeled by describing how specific parts of a UI are passed to a specific person, enabling him to manipulate specific shared data. In general, handling access to shared data and information provided by the underlying system supporting the multi-user dialog model also makes it possible to control collaboration to varying extents. Rights for reading and/or writing access to specific types of data can be defined in the model and again modified during runtime by supervisors or other participants, as long as they have reconfiguration rights for the model.

Collaboration can be handled by Petri net–based models, such as the turn-taking concept introduced by Mühlenbrock in [12]. Here, he describes the use of Petri nets for modeling the collaboration by defining whether a role is able to work on a shared task or not. This example of a Petri net–based model can be easily transferred to the reference net–based multi-user model.

The conceptual view of implementing CSCW systems using formally modeled MUIs introduced above is very general and needs further development. Still, we used a multi-user interface concept in a cooperative learning system for learning crypto-graphic algorithms [17]. Here, students were asked to cooperatively model a multi-user interface. They were asked to break up the protocol steps and allocate them to three different local UIs in a consistent way. The multi-user dialog model was prede-fined by a Petri net describing the correct protocol. Through cooperative modeling, the students were able to build a consistent cognitive model of the protocol and simu-late it using three individual (local) UIs.

This case study did not use the FILL-based approach described above, but could be easily translated because FILL is also transferred to reference nets. To do this, the initial step is to model the protocol in a cooperative way using FILL. Then, the global model is partitioned into several local interaction-logic models. For this, the students select those interaction elements from a toolbar that would enable them to launch the actions in the right context and at the right time according to their role in the protocol. Still, the protocol embedded in the reference net is used to generate error messages if any operations are distributed in a wrong way or have been executed in the wrong situation/context. Contextualized error messages are sent to the participant as long as this functionality has been introduced into the system process model by a teacher or supervisor under certain consistency checks. In the case of cryptographic protocols, it would be necessary to check whether the goal of the protocol is still reachable (e.g., exchanging keys) or not. In general, work has been done tackling the larger problem of synchronizing communication, as Paternò et al. show in [13], where they introduce a formal concept for synchronizing communication using CTTs.

6   Conclusions and Future Work

In this paper, we introduced a new concept allowing the modeler of a CSCW system to closely connect UI and cooperation modeling to adapt or change the cooperation model at runtime in response to changing requirements or tasks while reducing both conceptual errors and development costs. Here, the formal approach offers a chance to apply various validation and verification concepts to identify errors in the model or reconfiguration applied to it.

Future work mainly involves finalizing the implementation of the concept and conducting further evaluation studies to identify problems in the use of these concepts and detect technical problems that may arise when using the concepts in real-life scenarios. Finally, a great deal of work remains to be done on integrating existing CSCW concepts into the framework and exploring concrete reconfiguration operations that can be applied to existing multi-user interface models.

References

1. Arvola, M.: Interaction Design Patterns for Computers in Sociable Use. Int. J. of Computer Applications in Technology 25(2-3), 128–139 (2006)
2. Beauchemin, S., Barron, J.: Petri Net–based Cooperation in Multi-Agent Systems. In: Proc. of 2007 Conf. on Computer and Robot Vision, pp. 123–130. IEEE, New York (2007)
3. Borges, D., Neto, H., Neuman de Souza, J.: Work in Progress: Petri Nets as Applied to the Modeling of E-Learning Cooperative Systems. In: Proc. of 40th ASEE/IEEE Frontiers in Education Conf., pp. F1D-1–F1D-3. IEEE, Washington, DC (2010)
4. Bowers, J., Benford, S. (eds.): Studies in Computer-Supported Cooperative Work: Theory, Practice, and Design. Prentice Hall, New York (1991)
5. Ehrig, H., Hoffmann, K., Padberg, J.: Transformation of Petri Nets. ENTS 148, 151–172 (2006)
6. Furuta, R., Stotts, P.: Interpreted Collaboration Protocols and Their User in Groupware Prototyping. In: Proc. of 1994 ACM Conf. on CSCW, pp. 121–131. ACM, New York (1994)
7. Hao, Y., Huang, H., Zeng, P., Shao, W.: Collaborative Work Heterogeneous Data Sharing and Application based on Petri Net. J. of Adv. Mater. Res. 421, 732–738 (2012)
8. Johannsen, G.: Human-Machine Interfaces for Cooperative Work. Int. J. of Advances in Human Factors/Ergnonomics 20, 359–364 (1995)
9. Koch, M., Gross, T.: Computer-Supported Cooperative Work: Concepts and Trends. In: Proc. 11th IMS Conf., pp. 165–172. Koellen, Bonn (2006)
10. Kummer, O.: Referenznetze. Dissertation, University of Hamburg, Germany (2002)
11. Molina, A., Redondo, M., Ortega, M.: A Methodological Approach for User Interface Development of Collaborative Applications: A Case Study. J. Science of Computer Programming 74(9), 754–776 (2009)
12. Mühlenbrock, M.: Action-based Collaboration Analysis for Group Learning. Dissertation, University of Duisburg-Essen, Germany (2000)
13. Paternò, F., Mancini, C.: Developing Task Models from Informal Scenarios. In: Proceedings of CHI 1999, pp. 228–229. ACM, New York (1999)
14. UIMS Tool Workshop: A Metamodel for the Runtime Architecture of an Interactive System. SIGCHI Bulletin 24(1), 32–37 (1992)
15. Weyers, B., Burkolter, D., Luther, W., Kluge, A.: Formal Modeling and Reconfiguration of User Interfaces for Reduction of Errors in Failure Handling of Complex Systems. Int. J. of HCI (in press, 2012)
16. Weyers, B.: Reconfiguration of User Interface Models for Monitoring and Control of Human-Computer Systems. Dissertation, University of Duisburg-Essen, Germany (2012)
17. Weyers, B., Luther, W., Baloian, N.: Interface Creation and Redesign Techniques in Collaborative Learning Scenarios. Int. J. of Fut. Gen. Comp. Sys. 27(1), 127–138 (2011)

Using Collective Trust for Group Formation

Thomas Largillier and Julita Vassileva

MADMUC Lab
University of Saskatchewan
Saskatoon, Canada
{thomas.largillier,julita.vassileva}@usask.ca

Abstract. Group formation is a difficult task that arises in many different contexts. It is either done manually or using methods based on individual users' criteria. Users may not be willing to fill a profile or their profile may evolve with time without users updating it. A collaboration may also fail for personal reasons between users with compatible profiles as it may be a success between antagonist users that may start a productive conflict inside a team. Existing methods do not take into account previous successful or unsuccessful collaborations to forge new ones. The authors introduce a new model of collaborative trust to help select the "best" fitted group for a task. This paper also presents one heuristic to find the best possible group since in practice considering all the possibilities is hardly an option.

1 Introduction

There are many situations where people have to collaborate. It is an important job to make sure that the group gathered to accomplish a given task will perform efficiently. In learning context, it is often required that students perform exercises or projects as groups.

Several trust mechanisms have been developed over the years [1]. Most of those systems concern the trust a user A has in another user/product B. For example, Wang *et al.* in [11] define formally trust as "a peer's belief in another peer's capabilities, honesty and reliability *based on its own direct experiences*". They build a trust and reputation mechanism using Bayesian networks for file providers selection in peer-to-peer systems. Their approach helps users to find "better" peers in the system as well as even the load between file providers.

Gummadi *et al.* in [4] introduce a group to group trust value in peer-to-peer networks. However, their method forces all groups to be disjoint and the group-to-group trust between groups A and B is simply the average trust members of groups A have in members of group B. Therefore, this notion is simply an aggregation of trust collected in pairwise interactions.

Many virtual interactions nowadays are not between only two people, so there is a need to redefine trust metrics, since most of the existing ones always characterize the trust some user a has in one user/product b. Simply aggregating the pairwise trust will not help the user know which groups of users/products

V. Herskovic et al. (Eds.): CRIWG 2012, LNCS 7493, pp. 137–144, 2012.

she can trust. When interactions are group based, it is not enough to know that a peer is trustworthy, the user needs to know who he is trustworthy with and more importantly who he is not trustworthy with. A pairwise trust metric does not carry enough information. This paper introduces a collective trust mechanism together with an algorithm to compute the estimated trust of any group a user had no previous interaction with. This mechanism is then used to solve the group formation problem.

2   Group Formation

Several people addressed the problem of defining groups to perform a task. Most of the relevant work has been done in the field of collaborative learning and how to optimize the group formation phase so students will learn faster and better.

All the following approaches use individual characteristics of users to gather them into efficient groups.

Oakley *et al.* present in [10] their system to group students. Their team formation method aims at grouping together people with diverse ability levels with common blocks of time to meet outside classes. The groups are assembled by the teacher based on forms filled by the student. Each team member is also assigned a designed role inside the team. The roles change over time so that each student can see several aspects of team work. Since in this approach groups last at least a semester, the authors provide several guidelines on how to deal with problems like free riders in a group. The authors' scheme authorizes groups to be reshaped if a group wants to fire one of its members or if a group member wishes to leave her coworkers.

Martin *et al.* propose in [8] to use the Felder-Silverman [3] classification to adapt learning material to students as well as to group students in e-learning. The authors' idea is to gather both active and reflexive students inside groups to make the groups more efficient. Their idea, as well as the latent jigsaw method were used in class and described by Deibel in [2]. The feedback from the students was really positive as they say the groups help them to learn more efficiently and confronted them with new ideas.

Wessner *et al.* present in [13] a tool to group e-learning students. They introduce the Intended Points of Cooperation and how they can be used to form appropriate groups for a task. The grouping is done by hand by the teacher or can be done automatically to regroup people that have reached the same learning stage.

Inaba *et al.* in [5] propose to identify and describe users' personal objective using ontologies and to group people having similar objectives for collaboration to be more efficient. The collaborative learning ontology is developped further in [6] to provide a framework for group formation and designing collaborative learning sessions.

Muehlenbrock in [9] proposes severals ways to regroup people for efficient collaboration in learning. His system takes into account the users' availability detected automatically and also stores a static as well as a dynamic event profile for its users.

Wang *et al* in [12] propose a trust-based community formation method to recommend scientific papers. Users regroup around common interests and communities are built between users having a reciprocal high trust. Their method is based on a pairwise notion of trust and the trust a user has in a community is simply the average trust she has in its members.

All these methods use individual characteristics of the users and none of them uses the results from previous non-pairwise collaborations that may be really helpful in capturing all the complexity of human interactions. In the following section of this paper, we introduce a method based on the notion of collective trust and on the idea that groups that performed well in the past should perform well in the future. This method is orthogonal to all the methods presented in this section and can therefore be used to enhance the results provided by those as well as used alone.

3 Proposed Method

To overcome the limitations of the existing approaches, described in the previous section, we introduce the notion of collective trust. Having a non-pairwise trust metric allows to capture the interaction between users inside a group. For example, two users trusted independently can be untrustworthy when collaborating together and a small group can be really efficient while disappointing when integrated into a bigger structure. In reality personal factors may affect professional collaborations even if two people have compatible profiles. These notions are close to impossible to capture using individual profiles and have to be acquired with experience.

3.1 Collective Trust

This notion of collective trust is exactly the same as the trust defined in [11] except that it applies to groups of people/products instead of just one entity. It is based on the interactions someone has with a group of users/products.

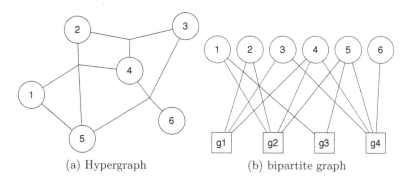

(a) Hypergraph   (b) bipartite graph

Fig. 1. Graph representations for collective trust

Let $\mathcal{U}$ be a set of users, for each $g \in 2^{\mathcal{U}}, Tr(g) \in [a, b] \cup \{\perp\}$. $Tr(g) = \perp$ means that g has never done a task and therefore has no trust value yet. Then after each interaction involving the group g, $Tr(g)$ is adjusted using the following formula:

$$Tr(g) = \begin{cases} (1 - \alpha) \cdot \frac{b-a}{2} + \alpha \cdot e & \text{if } Tr(g) = \perp \\ (1 - \alpha) \cdot Tr(g) + \alpha \cdot e & \text{otherwise} \end{cases} \tag{1}$$

where $\alpha \in [0, 1]$ is the learning rate and $e \in [a, b]$ is the result of the interaction valued on a scale from totally negative to totally positive.

This notion of trust can be represented by a bipartite graph or a hypergraph as shown on Fig. 1. In those representations the circle nodes represent the users. Every time a new group is assembled, an hyperedge (see Fig. 1a) or a group node (grey square, see Fig. 1b) is created or updated if it already exists. The two figures are equivalent and represent a state with 6 users where the four following groups $\{2, 3, 4\}, \{1, 2, 4, 5\}, \{1, 5\}, \{3, 4, 5, 6\}$ have already been put together at least once. The hyperedges/group nodes store all the group related information: trust, cardinality, number of interactions, etc. This way it is possible to access the information regarding the previous experiences of a user directly from her node.

3.2   Group Formation

The group formation problem consists in selecting the "best" group of people for one task, meaning the "group" that has the highest chance of success or that will outperform all the other possible groups on this particular task. The group formation problem can be modeled as follows:

$\mathcal{U} = \bigcup_{i=1}^{p} \mathcal{U}_i$ a set of users where $\mathcal{U}_i$ represents a specific type of users. The subset of available users is $\mathcal{U}_a \subseteq \mathcal{U}$. This notation naturally transpose to the types of users and $\mathcal{U}_{i_a}$ will denote the set of available users of type i.

A task $T = (t_i)_{1 \leq i \leq p}$ where $\forall i, t_i \in \mathbb{N}$ is a p-tuple specifying how many users of each type are required to accomplish the task. All tasks belong to the set $\mathcal{T}$.

A function $eval : \mathcal{T} \times 2^{\mathcal{U}} \to [a, b]$ that evaluates the success of a group on a specific task. Classic values for $[a, b]$ are $[0, 1]$ if the worst a group can do is being inefficient or $[-1, 1]$ if a group can worsen the situation by doing something.

The objective of the group formation problem is to find the group of available users $g \in 2^{\mathcal{U}_a}$ that fits the requirement of the task and that will maximize the $eval$ function over all the possible groups of available users. This can be written more formally as follows:

$$\begin{aligned} \mathcal{T} \times 2^{\mathcal{U}} &\longrightarrow & 2^{\mathcal{U}_a} \\ group : (T, \mathcal{U}_a) &\longrightarrow & group(T, \mathcal{U}_a) \end{aligned} \tag{2}$$

such as $\forall T \in \mathcal{T}, group(T, \mathcal{U}_a) = \emptyset \vee \forall i, |group(T, \mathcal{U}_a) \cap \mathcal{U}_i| = t_i$ that maximizes the value of $eval(T, group(T, \mathcal{U}_a)), \forall T$. This function either returns a group fit for the task or no group at all, if there are not enough available users of each type to complete the task.

We define $\mathcal{U}_a^T = \{u \in 2^{\mathcal{U}_a} | \forall i, |u \cap \mathcal{U}_i| = t_i\}$, i.e. this is the set of all possible groups for the task T. Then, $\forall g \in \mathcal{U}_a^T, |g| = \sum_{i=1}^{p} t_i = n$.

3.3   Collective Trust for Group Formation

Our method to find the "best" group for any given task is based on the collective trust metric introduced previously. The main idea is that collaborations that were efficient in the past should be efficient again, if put back together. The proposed *group* function is the following:

$$group(T, \mathcal{U}_a) = \mathrm{argmax}_{g \in \mathcal{U}_a^T} (ETr(g)) \tag{3}$$

where $ETr(g)$ is the estimated trust of the group g. The estimated trust of a group g is its trust value $Tr(g)$ if it has one. Otherwise, in order to estimate the trust we can have in a group that has never been put together before, we will look at the sub-groups it contains that have already been tested and use a linear combination of their weighted trust values as an estimate. The actual computation of the estimated trust goes as follows:

$$ETr(g) = \begin{cases} Tr(g) & \text{if } Tr(g) \neq \perp \\ \sum_{k=1}^{n} \frac{k}{\sum_{h \in C_g} |h \cap g|} \cdot \sum_{h \in C_g^k} \frac{k}{|h|} \cdot Tr(h) & \text{otherwise} \end{cases} \tag{4}$$

where $C_g^k = \{h \in 2^{\mathcal{U}} | Tr(h) \neq \perp \wedge |h \cap g| = k\}$ and $C_g = \bigcup_{k=1}^{n} C_g^k$. The idea is to use the trust of all groups $h \in C_g^k$ that share k members with the group g. $Tr(h)$ is multiplied by $k/|h|$ since only k members are selected and they only account for some amount of the whole group trust.

To guarantee that $ETr(g) \in [a, b]$, the value contributed by each group $h \in C_g^k$ is weighted by $k/(\sum_{h \in C_g} |h \cap g|)$. This particular weighting gives more importance to bigger groups since they will represent a bigger part of the final group and will have a bigger influence of the efficiency of the group.

Selecting the Best Group

Evaluating the estimated trust for a group can be done in linear time, regarding the number of users and groups, using either the bipartite graph or hypergraph representations depicted in Fig. 1. The computational problem comes from the number of possible groups g, for which trust needs to be estimated, for every task T:

$$|\mathcal{U}_a^T| = \prod_{i=1}^{p} \binom{t_i}{|\mathcal{U}_{i_a}|} \tag{5}$$

This number can grow really fast and become exponential which will be untractable in most cases. Therefore it is really important to consider approximate algorithms that will try to build the most trustworthy group for a task without actually computing all the estimated trusts.

Heuristic. This heuristic is a very simple greedy algorithm presented in Fig. 2. This algorithm builds a group by successively adding people from the most trusted groups. $\mathcal{G}$ represents the set of all groups with a non void trust value, *i.e.* $\mathcal{G} = \{g \in 2^{\mathcal{U}} | Tr(g) \neq \perp\}$. On step 3 of the algorithm, we select $i = n - |g|$ members at most from the available users. If the group we are selecting users

> *This algorithm returns the group corresponding to the task T with the highest estimated trust or the empty set.*
>
>   input: $\mathcal{U}_a$, available users
>   input: $\mathcal{G}$, preexisting groups
>   input: T, the given task
>   output: g, such that $g = \emptyset \vee |g| = n$.
>
> 1. $g = \emptyset$
> 2. **if** $\exists i, |\mathcal{U}_{i_a}| < t_i$ **then return** g
> 3. **while** $(|g| < n)$ **do**
> 4. $g' = argmax_{h \in \mathcal{G}}(Tr(h) \cdot \frac{i}{|h|})$ where $i \leq n - |g|$
> 5. $\mathcal{U}_a = \mathcal{U}_a \setminus g'$
> 6. $g = g \cup g'$
> 7. **return** g

Fig. 2. Greedy algorithm

from contains more than i users, we just select i users randomly. Steps 4 and 5 simply remove those users from the available ones and add them to the "best" group that will be selected for the task.

It is important to note that in the case where all the groups that have already been tested and possess a trust value are independent, the problem can be reformulated as a continuous 0-1 Knapsack problem [7] that is solved exactly by the greedy algorithm presented in Fig. 2.

4 Future Work

The first thing to do is to evaluate the proposed method trough extensive simulations and a real life experiment. The objective of the simulations will be to assess the quality of the proposed heuristic and to test its efficiency against several other methods like random assignment, methods presented in section 2 and pairwise trust schemes. The real life experiment will demonstrate the feasibility of the method.

A really important problem that requires further investigation is the estimated trust that one should have in a user that has never been part of any group. This estimated trust should be high enough to favor the incorporation of new users over members of poor previous collaborations but should not replace members of previous successful collaborations. The right threshold will be estimated using the simulations. It is important to notice that this threshold will in reality be task dependent. For example, it may be better to test new combinations on a common task while relying on known "good" teams for more critical tasks.

Another really important problem is the group partition problem. The objective here is not to find the "best" possible group to achieve a task but rather to separate the set of available users into groups of same sizes with homogeneous trust levels. This problem arises often in education where professors have to divide their classes for group work. If all the professors inside a university were to log the groups they made together with their performance, it will provide all the data required to compute the other classes' groups' estimated trust. Reusing the model presented in section 3, the group partition problem consists in finding a function:

$$\begin{aligned} \mathcal{T} \times 2^{\mathcal{U}} &\longrightarrow \text{Partition}(\mathcal{U}_a) \\ partition_\epsilon : (T, \mathcal{U}_a) &\longrightarrow (P_i^T)_i \end{aligned} \tag{6}$$

such as $\forall T, \forall i, \forall j, |P_i^T \cap \mathcal{U}_j| = t_j \wedge \forall T, \forall i, \forall j > i, |eval(T, P_i^T) - eval(T, P_j^T)| < \epsilon$

Our idea is to use the collective trust also for partitioning and make sure that all members of the partition have a similar estimated trust value. In order to provide a partition of $\mathcal{U}_a$ that provides groups with homogeneous trust levels, we will look for the partition that verifies one of the following properties:

$$\min_{P^T} \left(\sum_i \sum_{j>i} |ETr(P_i^T) - ETr(P_j^T)| \right) \tag{7}$$

$$\forall i, \forall j > i, |ETr(P_i^T) - ETr(P_j^T)| < \epsilon \tag{8}$$

Eq. 8 is more accurate since we want to guarantee that the level is homogeneous between groups but it might be difficult to set ϵ to get the best possible partition of users. On the other side Eq. 7 is always satisfied by at least one partition but this partition may not be really homogeneous especially if the number of groups in the partition is important. It is then application dependent to decide if having some outliers is really troublesome.

This collective trust metric can be adapted to recommend group of products to users. A good example can be online learning materials. People with different learning styles will be sensible to different kinds of learning materials and combination of learning materials.

5   Conclusion

In this paper, we presented a new scheme to select people to build a group based on the notion of collective trust. We strongly believe that this notion of collective trust is much more accurate in capturing the complexity of interactions between users than any individual based method. We also provided a heuristic to efficiently build the "most" trustworthy group. We will design simulations to prove the efficiency of the proposed heuristic as well as investigate other promising domains of application for the collective trust like recommendations of group of products.

References

1. Artz, D., Gil, Y.: A survey of trust in computer science and the semantic web. Web Semantics: Science, Services and Agents on the World Wide Web 5(2), 58–71 (2007)
2. Deibel, K.: Team formation methods for increasing interaction during in-class group work. In: Proceedings of the 10th Annual SIGCSE Conference on Innovation and Technology in Computer Science Education, ITiCSE 2005, pp. 291–295. ACM, New York (2005), http://doi.acm.org/10.1145/1067445.1067525
3. Felder, R., Silverman, L.: Learning and teaching styles in engineering education. Engineering Education 78(7), 674–681 (1988)
4. Gummadi, A., Yoon, J.: Modeling group trust for peer-to-peer access control. In: Proceedings of the 15th International Workshop on Database and Expert Systems Applications, pp. 971–978. IEEE (2004)
5. Inaba, A., Supnithi, T., Ikeda, M., Mizoguchi, R., Toyoda, J.: How Can We Form Effective Collaborative Learning Groups? In: Gauthier, G., VanLehn, K., Frasson, C. (eds.) ITS 2000. LNCS, vol. 1839, pp. 282–291. Springer, Heidelberg (2000), http://dx.doi.org/10.1007/3-540-45108-0_32,10.1007/3-540-45108-0_32
6. Isotani, S., Inaba, A., Ikeda, M., Mizoguchi, R.: An ontology engineering approach to the realization of theory-driven group formation. International Journal of Computer-Supported Collaborative Learning 4(4), 445–478 (2009)
7. Kellerer, H., Pferschy, U., Pisinger, D.: Knapsack problems. Springer (2004)
8. Martin, E., Paredes, P.: Using learning styles for dynamic group formation in adaptive collaborative hypermedia systems. In: Proceedings of the First International Workshop on Adaptive Hypermedia and Collaborative Web-based Systems (AHCW 2004), pp. 188–198 (2004), http://www.ii.uam.es/rcarro/AHCW04/MartinParedes.pdf
9. Muehlenbrock, M.: Learning Group Formation Based on Learner Profile and Context. International Journal on E-Learning 5(1), 19–24 (2006)
10. Oakley, B., Felder, R., Brent, R., Elhajj, I.: Turning student groups into effective teams. Journal of Student Centered Learning 2(1), 9–34 (2004)
11. Wang, Y., Vassileva, J.: Trust and reputation model in peer-to-peer networks. In: Proceedings of the Third International Conference on Peer-to-Peer Computing, P2P 2003, pp. 150–157. IEEE (2003)
12. Wang, Y., Vassileva, J.: Trust-based community formation in peer-to-peer file sharing networks. In: Proceedings of the 2004 IEEE/WIC/ACM International Conference on Web Intelligence, pp. 341–348. IEEE Computer Society (2004)
13. Wessner, M., Pfister, H.R.: Group formation in computer-supported collaborative learning. In: Proceedings of the 2001 International ACM SIGGROUP Conference on Supporting Group Work, GROUP 2001, pp. 24–31. ACM, New York (2001), http://doi.acm.org/10.1145/500286.500293

Time Series Analysis of Collaborative Activities

Irene-Angelica Chounta and Nikolaos Avouris

HCI Group, University of Patras, Greece
{houren,avouris}@upatras.gr

Abstract. Analysis of collaborative activities is a popular research area in CSCW and CSCL fields since it provides useful information for improving the quality and efficiency of collaborative activities. Prior research has focused on qualitative methods for evaluating collaboration while machine learning algorithms and logfile analysis have been proposed for post-assessment. In this paper we propose the use of time series analysis techniques in order to classify synchronous, collaborative learning activities. Time is an important aspect of collaboration, especially when it takes place synchronously, and can reveal the underlying group dynamics. Therefore time series analysis should be considered as an option when we wish to have a clear view of the process and final outcome of a collaborative activity. We argue that classification of collaborative activities based on time series will also reflect on their qualitative aspects. Collaborative sessions that share similar time series, will also share similar qualitative properties.

Keywords: time-series, collaboration, classification, logfile analysis.

1   Introduction

The analysis of collaborative activities is a complex task due to the nature of collaboration itself and the amount of information that has to be evaluated. However in computer supported collaboration, all the information regarding users' interaction is recorded in logfiles by the groupware applications that mediate such activities. Therefore logfile analysis, automated metrics and other quantitative methods are used for post-assessment of the quality of collaboration, to trace any possible drawbacks and reveal underlying mechanisms that may affect the process and outcome of collaborative activities [1-3]. Most of these methodologies take into account the aspect of time. For example, how turn taking mechanisms affect communication, whether large gaps in the communication flow might be considered as a failure or large periods of individual work phases might affect coordination [4]. We argue that such phenomena can be captured using time series [5]. In this study we use sequences of events of collaborative activities to form time series that represent how the process unfolds in time, related with quantitative assessments of collaboration quality. We explore whether collaborative sessions that share similar time series characteristics are also of similar quality. To that end, predictions of quality of collaboration based on time series techniques were compared to assessments by evaluators for a rich dataset of collaborative

V. Herskovic et al. (Eds.): CRIWG 2012, LNCS 7493, pp. 145–152, 2012.

activities. The use of time series as a tool of analysis of collaborative activities will add up and empower existing machine learning techniques while the workload of human evaluators will be minimized. Moreover, through time series techniques, real time assessment of the activity may be achieved. In that case, the evaluator will be aware, in real time, whether an activity is turning out successfully or, otherwise, which collaborative aspect should be further supported.

This paper is organized as follows. In section 2 the time series construction from the logfiles of various collaborative activities is described. In section 3 we discuss the structure and techniques used for the memory-based classification model that is proposed. The construction of the model itself is also analyzed. The results of the study are presented in section 4 and in section 5 we conclude with a general discussion about the setup and results of this study as well as improvements and future work.

2   Collaborative Activities as Described by Time Series

A computer-supported collaborative activity can be described via a multivariate time series of events (such as chat messages). We propose the use of multivariate time series because the way an activity builds up in time and its cross-correlation with other activity that occurs concurrently are important. To fully explore the underlying mechanisms and dynamics of collaboration such information might be proven useful and should not be ignored.

Time series is defined as any sequence of observations recorded at successive time intervals. Network traffic monitored by a web server per hour or the price of shares in a stock market per week are examples of time series commonly used and analyzed for various purposes. Time series fall into two categories: univariate and multivariate. A multivariate time series is a vector of more than one time series which are cross-related. The objective of time series analysis is to gain understanding of the nature and underlying mechanisms of a monitored activity, to group and classify samples based on their time series properties and to forecast. Many models and algorithms have been proposed to deal with univariate or multivariate time series analysis and classification, such as ARIMA, VAR, Hidden Markov Models (HMM), Dynamic Time Warping (DTW) and Recurrent Neural Networks. Time series analysis is widely used in a variety of fields such as economics, biology and computer science [6], [7].

For the purposes of this study we used the logfiles of collaborative activities that took place during a programming course in the Dept. of Electrical and Computer Engineering. The subject of the course was the joined construction of flow charts by dyads. The groupware application that mediated the activities provides users of a common workspace for the construction of diagrammatic representations and a chat tool to support communication between partners [8]. All activity was recorded in logfiles for later use. The setting of the study was such that same conditions applied for all clients/collaborators (e.g. equal numbered groups, high speed local area network, identical computers) and similar network delays occurred for all clients.

In order to portray a collaborative session as a time series we computed the sum of chat and workspace activity of collaborating partners per time intervals and per events. The sequences of aggregated chat and workspace events form up a multivariate time series per collaborative session. Previous studies show that a number of metrics regarding chat and workspace activity highly correlate with the quality of collaboration of a joined activity [9]. Based on these studies we made use of the metrics displayed in Table 1 to assemble time series for each session. Therefore each collaborative session is represented by one multivariate time series which is practically a vector of four univariate time series constructed from aggregated chat and workspace events, where:

— Number of chat/workspace actions per time interval: the sum of messages/workspace actions of both partners in a time interval.
— Roles' alternations in chat activity per time interval: the number of times the active role of a partner was switched in chat/workspace activity in a time interval.

Table 1. Chat and Workspace metrics used for time series construction

Metric	Sums	Difference of Sums	Alternations	Difference of Alternations
chat	number of chat messages per time interval	difference of chat messages between consecutive time intervals	roles' alternations in chat activity per time interval	difference of roles' alternations in chat activity between consecutive time intervals
workspace	number of workspace actions per time interval	difference of workspace events between consecutive time intervals	roles' alternations in workspace activity per time interval	difference of roles' alternations in workspace activity between consecutive time intervals

Another critical point when creating time series from aggregated events is the time interval chosen [10]. Valuable information might be lost in case of a small or large time interval. The choice of the appropriate time interval is a critical task and domain dependent. Therefore a variety of time intervals was tested before concluding. In this study, the duration of collaborative activities ranged between 50 minutes to 1 hour and a half. The time intervals studied were of 1, 5, 8 and 10 minutes. Activity for time intervals of less than one minute was not explored since the number of events occurring within such time periods was small.

3   Methodology of Analysis

The aim of this study is the classification of collaborative sessions using their time series properties. For this purpose we create a data pool of time series, extracted from collaborative activities, associated with quantitative assessments of collaboration quality. The suggested set up is fashioned after memory-based learning models using time series [11]. Memory-based learning presupposes the use of a distance measure. For that we used the Dynamic Time Warping (DTW) algorithm.

3.1   Dynamic Time Warping

DTW is an algorithm that measures similarity between two sequences that vary in time. It provides a distance measure termed DTW distance. Originally it was used for sound and video processing but has also found many applications in time series analysis [12], [13]. We used the DTW algorithm implementation proposed by Giorgino, T., for the R-statistics software [14]. This implementation allows choosing between multiple options for step patterns (the way consecutive time series elements are matched) and dissimilarity functions (for the cross-distance matrix computation). The DTW algorithm does not presuppose time series' stationarity or non-missing information as Fourier transform does and this is one of the reasons for its growing popularity.

3.2   Quality of Collaboration

The rating scheme proposed by Kahrimanis et al. [3] was used to provide quantitative judgments of the quality of collaboration for the sessions used in this study. The aforementioned rating scheme proposed the rating of seven collaborative dimensions on a 5 point scale, that stand for the five, fundamental aspects of collaboration: communication, joint information processing, coordination, interpersonal relationship and motivation [4]. These seven dimensions are: collaboration flow, sustaining mutual understanding, knowledge exchange, argumentation, structuring the problem solving process, cooperative orientation and individual task orientation. The rating was carried out by two trained evaluators. We made use of the average value of six out of seven, dimensions leaving out the motivational/Individual task orientation aspect which is rated for each student separately. We denote this metric as Collaboration Quality Average (CQA) and it takes values within the range {-2,2}. As stated in Section 2, CQA has been found highly correlated with logfile metrics of interaction. Therefore we argue that similar time series will have similar CQA evaluative values.

3.3   Memory-Based Classification Model

The memory model construction and classification procedure consists of three steps.

1. Time series construction from the logfiles of collaborative sessions. For each collaborative session we constructed its multivariate time series representation, as described in section 2. Outliers were detected by visual inspection of time plots and deleted from the final dataset.
2. Input of sample entries in memory. The data pool consists of the multivariate time series extracted from collaborative activities, 212 samples in total, as collected in step 1. Each sample's quality of collaboration is also assessed by two evaluators (section 3.2). Therefore each point in the memory stands for a collaborative session and is described by its time series and an evaluation value for the quality of collaboration (CQA).
3. Classification of a query sample. By the term "query sample" we name any collaborative session that is not accompanied by an evaluation value CQA. The purpose is to approximately estimate the evaluation value CQA by finding the optimal time

series match of the session among the sessions of the data pool. We argue that collaborative activities described by similar time series will have a similar CQA evaluation value. Therefore if the time series of two samples ts_Sa and ts_Sb, with Sa being the query sample and Sb the reference sample, have a minimum distance DTW then the evaluative value CQA of Sa should be approximately equal to the evaluative value CQA of Sb.

4   Results

The dataset used initially consisted of 228 collaborative sessions. The logfiles of the sessions, as recorded by the groupware application that mediated the activity, were used for the construction of the multivariate times series of aggregated events. Outliers were removed and the final dataset used in the memory-based classification model consisted of 212 collaborative sessions. For each one of the 212 samples we computed the DTW distance and found the optimal match from the 211 samples remaining in the data pool. The study was repeated for a variety of time intervals in aggregated events (1, 5, 8 and 10 minutes), two dissimilarity functions (Euclidean and Manhattan) and two step patterns (symmetric1 and symmetric2) of the DTW algorithm. In order to evaluate the results, as well as define the most appropriate time interval, dissimilarity function and step pattern, we estimated the correlation matrix of the evaluative value CQA (predicted vs. true value), the root mean squared error (RMSE) and the mean absolute error (MAE) for each case.

Correlation is a popular method to explore statistical relations between variables. Spearman's rank correlation coefficient is used to reveal any existing relations between the evaluative value CQA of a session, as assessed in the evaluation phase, and the predicted - by the memory-based classification model - value. For step pattern set to the symmetric $P = 0$, the two variables are significantly correlated for most of the combinations of time intervals and dissimilarity methods. Spearman's rho depicts the degree of the relation between two variables and it may range from -1 to 1. A value of 1 shows a strong, positive correlation while a value of -1 reveals a strong, negative. In our case the real and predicted evaluative values of CQA are positively and significantly correlated for all time intervals (Table 2). However the strongest correlation occurs for 1 minute time interval and Manhattan as a dissimilarity method ($p<0.05$, rho=0.3).

Table 2. Spearman's Rho correlation coefficient for CQA real and predicted values per time interval, for Manhattan and Euclidean dissimilarity methods and for step pattern symmetric P=0

	Manhattan		Euclidean	
time interval	p value	Rho	p value	Rho
1 minute	0,000	0,3	0,029	0,15
5 minutes	0,002	0,2	0,021	0,15
8 minutes	0,000	0,23	0,005	0,18
10 minutes	0,011	0,17	0,010	0,17

The mean absolute error (MAE) and root mean squared error (RMSE) are used to measure accuracy; how close the predicted values are to the "observed", or real, ones. For MAE individual differences are weighted equally in the average while large errors are highly weighted in RMSE. The difference of MAE and RMSE can be used as an insight of the variance of individual errors in the sample. We provide the results of analysis for both error metrics in Table 3. Smallest MAE and RMSE values occurred for 1 minute time interval and Manhattan method (MAE=0.89, RMSE=1.1) while largest values occurred for 5 minutes interval and Euclidean distance (MAE=1.21 RMSE=1.5). The variance of individual errors is also minimized for 1 minute time interval. The extent to which the alignment of DTW avoids mismatches is influenced by the choice of the dissimilarity function (Euclidean or Manhattan) [14]. Therefore the difference portrayed in the results can be justified.

Table 3. MAE and RMSE for CQA real and predicted values per time interval, for Manhattan and Euclidean dissimilarity methods and for step pattern symmetric P=0

Time Interval	MAE		RMSE		RMSE - MAE	
	Manhattan	Euclidean	Manhattan	Euclidean	Manhattan	Euclidean
1min	0.89	0.97	1.14	1.21	0.25	0.24
5min	1.19	1.21	1.48	1.50	0.29	0.29
8min	1.18	1.16	1.50	1.48	0.32	0.32
10min	1.17	1.19	1.44	1.47	0.27	0.28

Both correlation and error measures reveal that time series of aggregated events per time interval of 1 minute portray better the quality of collaboration of a group activity as this has been assessed by human evaluators using a rating scheme. In case we use the same metrics for larger time intervals of the activity, valuable information might be lost (Fig. 1). For time interval of 1 minute, the distribution of absolute difference among the CQA predicted and CQA real value is shown in Fig. 2. The absolute difference for 41% of cases was less than 0.5, for 68.4% of the cases difference was less than 1 while the 92% falls below a difference of 2 points.

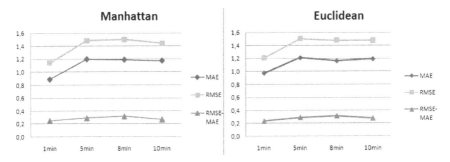

Fig. 1. MAE and RMSE per dissimilarity function

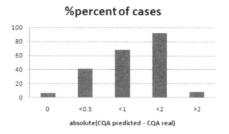

Fig. 2. Distribution of absolute difference among the CQA predicted and CQA real values

5   Conclusions and Future Work

In this paper we discuss the use of time series for classifying collaborative activities. The main goal of the study was to explore whether this classification regarding the time series characteristics of collaborative session depict also their qualitative assessments of collaboration quality. For the classification a memory-based approach was proposed and the Dynamic Time Warping algorithm was used as a dissimilarity measure between time series. The results revealed that there is a significant positive correlation among the predicted and real evaluative values (CQA) while the lower values for MAE and RMSE values as well as for the variance of individual errors in the sample occur for 1 minute time interval (0.89, 1.1 and 0.25 respectively, for a value range $\{-2, 2\}$). The error metrics MAE and RMSE are used to provide insights of a model's accuracy. In order to have a more complete picture we should also take into account qualitative aspects such as the simplicity of the model and the context within it is used.

In the classification procedure one optimal match was used for each query time series. Results could be improved if we used more advanced techniques such as k-nearest neighbor or applied weights in the DTW algorithm. Moreover other classification models for multivariate time series should be studied, such as Hidden Markov Models [15], as well as popular clustering methods like hierarchical clustering and kmeans. In future work, a qualitative analysis regarding time series characteristics and how they are related with individual aspects of collaboration will be explored, as well as how random shocks, such as communication failure due to network breakdowns, affect the collaborative process and the way partners recover from that.

References

1. Barros, B., Verdejo, M.F.: Analysing student interaction processes in order to improve collaboration. The DEGREE approach. International Journal of Artificial Intelligence in Education 11(3), 221–241 (2000)
2. Schümmer, T., Strijbos, J.-W., Berkel, T.: A new direction for log file analysis in CSCL: Experiences with a spatio-temporal metric. In: Koschmann, T., Suthers, D., Chan, T.W. (eds.) Computer Supported Collaborative Learning 2005: The next 10 years!, pp. 567–576 (2005)

3. Kahrimanis, G., Meier, A., Chounta, I.-A., Voyiatzaki, E., Spada, H., Rummel, N., Avouris, N.: Assessing Collaboration Quality in Synchronous CSCL Problem-Solving Activities: Adaptation and Empirical Evaluation of a Rating Scheme. In: Cress, U., Dimitrova, V., Specht, M. (eds.) EC-TEL 2009. LNCS, vol. 5794, pp. 267–272. Springer, Heidelberg (2009)

4. Meier, A., Spada, H., Rummel, N.: A rating scheme for assessing the quality of computer-supported collaboration processes. International Journal of Computer-Supported Collaborative Learning 2(1), 63–86 (2007)

5. Vasileiadou, E.: Stabilisation operationalised: Using time series analysis to understand the dynamics of research collaboration. Journal of Informetrics 3(1), 36–48 (2009)

6. Box, G.E.P., Jenkins, G.M., Reinsel, G.C.: Time series analysis, Forecasting and Control, 4th edn. Wiley (2008)

7. Chatfield, C.: The Analysis of Time Series: An Introduction. Chapman & Hall (2003)

8. Avouris, N., Margaritis, M., Komis, V.: Modelling interaction during small-group synchronous problem-solving activities: The Synergo approach. In: 2nd Int. Workshop on Designing Computational Models of Collaborative Learning Interaction, ITS 2004, Maceio (2004)

9. Kahrimanis, G., Chounta, I.-A., Avouris, N.: Study of correlations between logfile-based metrics of interaction and the quality of synchronous collaboration. International Reports on Socio-Informatics (IRSI) 7(1), 24–31 (2010)

10. Shellman, S.M.: Time series intervals and statistical inference: The effects of temporal aggregation on event data analysis. Political Analysis 12(1), 97–104 (2004)

11. Deng, K., Moore, A., Nechyba, M.: Learning to recognize time series: Combining arma models with memory-based learning. In: IEEE CIRA, pp. 246–250 (1997)

12. Warren Liao, T.: Clustering of time series data—a survey. Pattern Recognition 38(11), 1857–1874 (2005)

13. Chao, S., Wong, F., Lam, H.-L., Vai, M.-I.: Blind biosignal classification framework based on DTW algorithm. In: International Conference on Machine Learning and Cybernetics (ICMLC), vol. 4, pp. 1684–1689 (2011)

14. Giorgino, T.: Computing and visualizing dynamic time warping alignments in R: the dtw package. Journal of Statistical Software 31(7), 1–24 (2009)

15. Kirshner, S.: Modeling of multivariate time series using hidden Markov models. PhD thesis, University of California (2005)

SoCCR – Optimistic Concurrency Control for the Web-Based Collaborative Framework Metafora

Andreas Harrer, Thomas Irgang, Norbert Sattes, and Kerstin Pfahler

Catholic University Eichstätt-Ingolstadt, 85072 Eichstätt, Germany
{andreas.harrer,thomas.irgang,norbert.sattes,
kerstin.pfahler}@ku-eichstaett.de

Abstract. In this paper we present the concurrency control used in the computer-supported collaborative learning framework Metafora. Metafora is an environment that supports complex learning scenarios utilizing multiple learning tools, such as a tool for the planning of learning activities, a graphical argumentation tool and several microworlds in the domains of science and mathematics. Since Metafora is a web-based framework, specific requirements have to be fulfilled for smooth collaboration and inter-tool communication. For smooth collaboration we will describe our optimistic concurrency control approach that allows concurrent modification of shared objects in a workspace as far as possible. While move and edit actions can be performed in parallel, a **S**ocial **C**oncurrency **C**onflict **R**esolution (SoCCR) protocol enables collaborative editing of text nodes in the planning space. We will illustrate this with an example of user interaction in the Metafora system involving the concurrency mechanism.

Keywords: Web-based collaborative applications, collaborative workspaces, computer-supported collaborative learning (CSCL), concurrency control.

1 Introduction - Web-Based Collaboration in Metafora

Metafora is an ongoing European project[1] that combines different pedagogical strands, namely constructionism and collaboration, resulting in an approach called *learning to learn together* (L2L2). Constructionism [HP91] stresses an active role of the learner who is (re-)constructing knowledge by herself instead of knowledge being delivered by the teacher. Usually this is achieved by direct construction of artefacts, models, programs etc. Collaboration is another facet to engage students to a more active attitude during learning, stimulating argumentation, negotiation, planning and different kinds of strategic skills referring to management and task solution. Metafora brings together a set of different

---

[1] The Metafora project is co-funded by the European Union under the Information and Communication Technologies (ICT) theme of the 7th Framework Programme for R&D (FP7), Contract No. 257872, http://www.metafora-project.org/

V. Herskovic et al. (Eds.): CRIWG 2012, LNCS 7493, pp. 153–160, 2012.

learning tools from science and math within a framework for collaborative and self-regulated learning and organization of the learning process. Among these tools are so called Microworlds for Math and Physics, game-like environments for sustainability and ballistics, and editors for the construction of mathematical patterns and algebraic equations. This is combined with Metafora's general features for collaboration via a planning space, a group chat and the LASAD discussion environment[2]. Metafora is designed as a web-based platform embedding and connecting multiple learning tools and allows seamless transition and exchange of information between these tools (see figure 1).

Fig. 1. Metafora system with core functionality (login + group management, chat and toolbar) on the left and the physics microworld 3DJuggler visible in the main part

In contrast to regular internet-based collaborative systems where bidirectional communication is possible between different remote instances of the collaborative application, web-based collaboration has to overcome one deficit: classical web-based interaction follows the request–response–schema, i.e. a web client issues a request to a web server and the server sends a response back. While this can be used in collaboration to bring a user's action from the web client to the server, all the other users' web clients have to get notice of the same action. One solution is to weaken the strict request–response schema and use techniques called "server push" that allow the server to actively send messages to web clients. A family of methods for this is known under the name "comet"[3] and has been used in collaborative web applications. Metafora also uses such a library for the built-in chat, the graphical *Planning Tool*, and the propagation of updates in the awareness and sharing tool *Workbench*.

---

[2] http://cscwlab.in.tu-clausthal.de/lasad

[3] http://code.google.com/p/google-web-toolkit-incubator/wiki/ServerPushFAQ

2   Concurrency Control in the Shared Planning Space

The Planning Tool is a web-based application that provides a visual language for planning, enactment, and reflection of Metafora learning activities. This language consists of *cards* (boxes) for activities, methods to realise activities, roles, attitudes, and *edges* / connectors between these cards. Cards hold a textfield where users can comment and explain the meaning of it to their peers. Even though it is built as a stand-alone web application it unfolds its full potential embedded into the Metafora platform and connected to the other learning tools. It is possible to create plans for conducting a challenge / inquiry, and to directly enter tools from this plan seamlessly, using automatic login to the other tools and providing the work context needed to tackle specific tasks within the challenge. The plan can also be used as a documentation tool, by checking / unchecking activities as started and finished, so that the students can use the plan as a graphical organizer of their achievements. Finally, the plan can be considered a living document that is constantly being revised, checked and taken as an artefact for reflection about students own work and organization. An example plan can be found in figure 2.

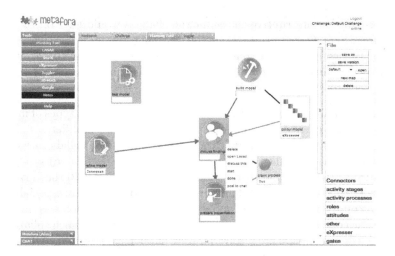

Fig. 2. The Planning Tool with a plan in use, showing started cards coloured in yellow, finished cards in green, and all cards having a text area to add notes

Technically, the planning tool uses the Google Web Toolkit for the graphical user interfaces and the client-server communication to transmit user actions performed in a web browser. Since it allows collaborative usage remotely, actions being performed by one user have to be propagated actively to all the peer students working in the same context / group. For this end, we use a server push technology[4] as explained in the previous section to overcome the limitations of the conventional request-response protocol of web-based applications.

---

[4] http://code.google.com/p/gwt-comet/

2.1   Formal Analysis of Concurrent Actions

As for concurrent actions performed by different users, not each combination of typical actions has to be considered for conflicts. While in principle the set of actions in the planning space consists of createCard, modifyCard, deleteCard, createEdge, and deleteEdge, some of them have a clear effect regardless of being executed concurrently or isolated: any combination involving deleteCard will lead to the deletion of the planning card, while creation of a card cannot conflict with any manipulation of it as long as the creation is not finished. Similarly this holds for creation and deletion of edges. Thus, as the interesting aspect remains the modification of nodes / cards, which has two different facets:

Move (m) is the operation that changes the position of a card in the planning space, i.e. the X- and Y-coordinates of the objects will change. Even though dragging the object might take several seconds, this action is usually relatively short.

Edit (e) is the operation of creating and modifying text in the text area of a card in the planning space. In our implementation the changes are not delivered and synchronized per character as in GROVE [EG89] or GoogleDocs, because of the large number of messages and thus server load produced and because of the interruptive character the changes would have on the receiving users' side. Our approach is similar to a chat system, where messages are typed and committed by sending them via pressing the return key. Given this, our variant of text production is a relatively time-consuming operation and prone to conflicts.

Because of the different average duration of performing the two types of modifications the handling of concurrency should take into account this difference. Locking the object [BS00] during any modification would inhibit the collaborative usage, especially in the case of an edit which might lock the respective object for a long period of time. Similarly, approaches like floor passing [KAF02] that give exclusive control to one user at a time hamper a smooth collaboration, especially when considering our project purpose of supporting L2L2. Our decision was then to use an optimistic concurrency control approach[BS00] allowing as much collaboration as possible where conflicts are resolved if they happen. The formal analysis of the effects of concurrent moves and edits by two or more users is presented in table 1.

Table 1. Possible concurrent actions and results for two users U_1 and U_2

$U_1 \backslash U_2$	Move	Edit
Move	$m \in \{m_1, m_2\}$	$m_1 \circ e_2$
Edit	$m_2 \circ e_1$	Conflict

When two users concurrently move a node, the resulting position is the position of the user action which reaches the server last. To avoid interleaving between sending the action and receiving a new position, which might result in both clients receiving the other users' modified position after sending, each action is echoed / resent also to the original sending client. This guarantees a consistent result on both clients.

Concurrent move and edit do not conflict with each other, because different object attributes are modified, thus allowing arbitrary ordering of the actions, i.e. commutativity. This results in both cases in the formal composition of move followed by the edit action. For the user typing text, the move performed by the other user is visible on the screen but not interrupting the text production.

Concurrent editing is the most critical case to consider: since both actions usually take a long time, concurrent usage is the most likely combination in the table. Locking the whole card would hinder other users to act for a long time, while it is well possible that a user edits the text without commiting the changes, resulting in locking others out without need. Our decision is to allow concurrent editing and only in case of the commit of text changes a specific mechanism for **S**ocial **C**oncurrency **C**onflict **R**esolution (SoCCR) comes into action. The user that receives commited changes from the other user will probably have a different text content than the collaborator's text. In this case the receiving user gets prompted that there is a conflict with the options of choosing the peer's text, his / her own text or an integrated version of both. This approach is similar to the mechanism wikis (or code versioning systems) use in case of edit conflicts [SF10], yet in the granularity of card texts instead of whole pages in the wiki.

2.2 Implementation of Concurrency and SoCCR Conflict Resolution

The implementation of our proposal from the previous subsection makes use of three variables for the text content of the concurrently changed card at each client's side. These variables respresent the following aspects:

U Uncommited local is the content of the text area at the current moment
C Commited local is the latest commited text, i.e. the text before editing started
N New remote is the content that has been committed at a different client's side and has been propagated by the server to this client recently

Not every combination of (in)equality of the variables is possible, e.g. $U = C, C = N, N \neq U$ is impossible because of transitivity. This reduces the number of constellations to 5, as in table 2.

The two cases of $U = C$ are not critical: there is no uncommitted local edit, so that the remote action can be performed without problems. Given $C = N$ it is a remote Move (m_r) and for $C \neq N$ it is a remote Edit (e_r).

In contrast the constellations with $U \neq C$ have an uncommited local edit:

Table 2. Constellations and effects for variables U, C, N

Local		Remote		Effect
$U = C$	no local edit	$C = N$	remote move	perform move
$U = C$	no local edit	$C \neq N$	remote edit	update U and C
$U \neq C$	local edit	$C = N, U \neq N$	remote move	perform move
$U \neq C$	local edit	$C \neq N, U = N$	remote edit	no conflict, update C
$U \neq C$	local edit	$C \neq N, U \neq N$	remote edit	**show SoCCR conflict dialog**

- $C = N$ means a remote move (m_r) that can be performed while keeping the uncommited text in the card, but allowing the concurrent position change.
- $C \neq N$ and $U = N$ means the remotely commited text equals the local uncommited, thus allowing to change C to N, yet without any conflict to U.
- $C \neq N$ and $U \neq N$ represent a remote edit commited that is neither compatible with the current text nor the last local. This means that committing the local change would produce a conflict, because it would overwrite a text the user didn't see and should be made aware of.

We will illustrate our current implementation with screenshots showing the user interface of this constellation. Figure 3 shows two users Alice and Nobbi concurrently editing the same card. Locally uncommitted changes are visualised with a light red background. Alice and Nobbi entered different opinions about how many iterations of testing should be performed (one respectively five).

After one of the users – here Nobbi – commits his change (this changes the background colour at his client's side to a light green to show successful change, see top of figure 4), exactly the situation that commited local (C), current local (U) and newly commited remote (N) text (from Nobb)i are different at Alice's client. Our solution is to prompt the "local user" with a conflict dialogue, that offers all the information needed to resolve this conflict: both conflicting texts N

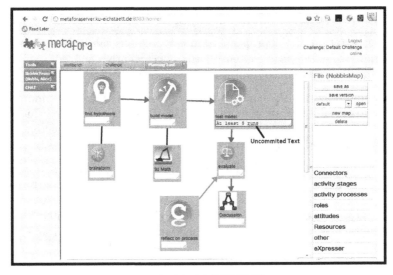

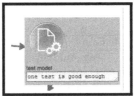

Fig. 3. Screenshot of two users Nobbi (top) and Alice (bottom) editing concurrently

and U are shown and a textbox is offered where the user should solve the conflict by filling in the text she wants to confirm. This could be either option N, U, the continuation of U according to the users original intention, or a completely different text that takes into account both positions. To stimulate reflection about the texts the initial content of the textbox is the text of the remote user, not the own which could be commited without any further thought. In our example visible in figure 4 Alice chooses to integrate both opinions, i.e. five or just one test run, into a compromise of three test runs.

This approach is also usable in collaboration groups with concurrent editing of more than two users. Conflicts are shown as presented and are being updated continuously if another edit is commited while the conflict dialogue is open the users are updated about these changes, too.

The elaborated analysis of the different cases shows that most concurrent modifications can be handled technically well without any locking, thus allowing more collaboration than with the pessimistic concurrency control approaches. The one critical case of edit conflict is supported by making the user aware of a textual conflict and offering a user interface to resolve this conflict. Thus we provide means to stimulate a social protocol and mediation on the learners' side instead of a techni-

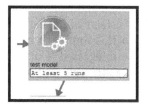

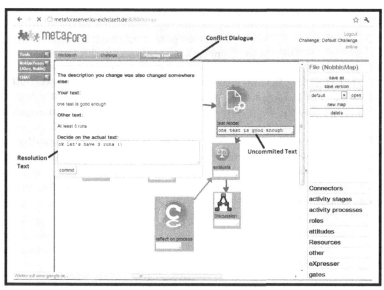

Fig. 4. Screenshot after user Nobbi (top) commited, Alice (bottom) gets a conflict dialogue

cal protocol. We believe that this solution supports the spirit of our L2L2 concept better, because agreement and reflection on the conflict solution are encouraged. We plan to compare the flow of interaction and user satisfaction with regular lock-based or floor-passing collaboration in a controlled lab study.

3   Conclusion and Further Work

In this paper we presented the web-based collaboration framework Metafora and discussed specifically the concurrent text editing of shared text cards in the Planning Tool. In order to allow a high degree of collaboration on the same objects we decided against a lock-based, turn taking or floor-passing approach, which all reduce the potential for concurrent manipulation. This is especially true for text editing, which is a time consuming operation, where a lock would hinder other users from contributing for a long time. Our solution uses an optimistic approach that allows concurrent manipulation and indicates conflicts in case of text changes in the same object. Our implementation attempts to stimulate users' reflection about their texts and foster integration of the different positions. This is based on our principle of promoting "learning to learn together" (L2L2). Thus, we deliberately support and induce a Social protocol for Concurrency Conflict Resolution (SoCCR) instead of letting a technical protocol guide the learners. Possible extensions to this approach are to involve the remote committer in the negotiation of the integrated text by also prompting him / her a conflict dialogue or to use voting strategies [BS00] for the conflict resolution.

References

BS00.   Borghoff, U.M., Schlichter, J.H.: Computer-Supported Cooperative Work: Introduction to Distributed Applications. Springer, Berlin (2000)

EG89.   Ellis, C.A., Gibbs, S.J.: Concurrency control in groupware systems. In: Clifford, J., Lindsay, B., Meier, D. (eds.) Proceedings of the 1989 ACM SIGMOD International Conference on Management of Data, pp. 399–407. ACM (1989)

HP91.   Harel, I., Papert, S. (eds.): Constructionism. Ablex Publishing, Norwood (1991)

KAF02. Komis, V., Avouris, N., Fidas, C.: Computer supported collaborative concept mapping: Study of interaction. Education and Information Technologies 7(2), 169–188 (2002)

SF10.   Shelly, G.B., Frydenberg, M.: Web 2.0: Concepts and Applications. Cengage Learning / Course Technology, Boston (2010)

Enabling and Evaluating Mobile Learning Scenarios with Multiple Input Channels

Lars Bollen[1], Sabrina C. Eimler[2], Marc Jansen[3], and Jan Engler[2]

[1] University of Twente, Dept. of Instructional Technology, The Netherlands
l.bollen@utwente.nl
[2] University of Duisburg-Essen, Dept. of Computer Science
and Applied Cognitive Science, Germany
sabrina.eimler@uni-due.de, engler@collide.info
[3] University of Applied Sciences Ruhr West, Computer Science Institute, Germany
marc.jansen@hs-ruhrwest.de

Abstract. Applications and research efforts in Mobile Learning constitute a growing field in the area of Technology Enhanced Learning. However, despite a permanent increase of mobile internet accessibility and availability of mobile devices over the past years, a mobile learning environment that is easy to use, widely accepted by teachers and learners, uses widespread off-the-shelf software, and that covers various application scenarios and mobile devices, is not yet available. In this paper, we address this issue by presenting an approach and technical framework called "Mobile Contributions" ("MoCo"). MoCo supports learners to create and send contributions through various channels (including third-party solutions like Twitter, SMS and Facebook), which are collected and stored in a central repository for processing, filtering and visualization on a shared display. A set of different learning and teaching scenarios that can be realized with MoCo are described along with first experiences and insights gained from qualitative and quantitative evaluation.

Keywords: mobile learning, heterogeneous devices, multiple input channels, SMS, Twitter, Facebook, visualization, one-minute paper, self-learning phases, evaluation.

1 Introduction

Mobile internet usage has been on the rise in the past years, and recent studies indicate that this is an ongoing trend for the coming years [1]. Even more, mobile internet usage is expected to surpass desktop internet usage. Similarly, private ownership of devices that can be used for mobile internet access became more and more common. Mobile phones, smartphones, tablets, and notebooks, are in close reach of most learners nowadays. The combination of an increasing accessibility to the internet and commonly owned mobile devices of some sort paved the way of applying these technologies in learning and teaching. Research and development in the area of Mobile Learning advanced significantly in the past two decades as pointed out in a recent overview article by Kukulska Hulme [2]. Many research activities and application can

V. Herskovic et al. (Eds.): CRIWG 2012, LNCS 7493, pp. 161–175, 2012.

be found on mobile language learning, field trip support, classroom response systems, discussion support systems etc. However, largely, these environments and applications are only used and applied in research contexts and experiments, and rarely reach everyday school life for several reasons:

— Mobile phones (or their usage, resp.) are banned in most schools. On an institutional level, mobile devices are considered a nuisance or even a threat to established learning activities.
— Mobile learning activities are more difficult to control and require open-minded, daring teachers.
— Mobile learning systems are difficult to set up, to configure and maintain. School administrators shun the effort to manage non-standard technologies.
— Many (mobile) learning environments focus on one kind of activity (e.g. a classroom response system) or on one domain (e.g. butterfly-watching [3]), thus limiting potential uses.
— Schools cannot afford to buy large, homogeneous sets of mobile devices, but many mobile learning applications aim at special devices or operating system.

We argue that to reach large numbers of learners, teachers, and schools, a mobile learning environment has to

— work with all commonly owned computational, mobile and non-mobile devices,
— be universal, generic and flexible to support various educational scenarios,
— be easy and intuitive in usage,
— be regarded useful and bring an added value to teachers,
— bring opportunities for research about the use, acceptance and effect of incorporating students' contribution from various (mobile device) channels in everyday learning and teaching.

In this paper, we describe an approach called "Mobile Contributions" ("MoCo") that addresses the mentioned shortcomings of existing mobile learning scenarios and is anticipated to reach large numbers of learners and teachers in academia and schools. At the core, MoCo allows learners to create textual contributions (questions, comments, answers, etc.) by using arbitrary devices and communication channels, e.g. SMS, Twitter, Facebook, e-mail or a web page. Contributions are abstracted and aggregated for storage and retrieval, while a visualization component provides features for filtering, reviewing and presenting contributions in various educational scenarios.

With MoCo, we try to utilize commonly and widely used communication channels that are frequently used by learners and teachers alike in an integrated way to put an innovative, mobile learning environment into practice. Being highly generic and flexible, i.e. not relying on certain devices or content domains, and not enforcing predefined activity structures, MoCo is intended to be suited to realize a large number of different scenarios, e.g. brainstorming activities, classroom discussion support, field-trip support, one-minute papers [4, 5], self-learning phases etc.

As an ultimate objective, the presented approach aims at supporting a large number of different devices (that only need to be able to communicate via SMS or have internet access) to realize different mobile learning scenarios. These scenarios cover a large number of categories, e.g. synchronous and asynchronous scenarios, formal and informal learning scenarios, individual or group work etc. In addition, smooth transitions between those categories shall be possible to tear down the borders between different learning settings for teachers, moderators and learners.

In the following section, we will relate our approach to a number of typical mobile learning systems that bear some resemblance to MoCo. In section 3, we describe the technical design and first implementations of our system. Section 4 will describe a number of application scenarios along with first experiences and evaluations. Finally, section 5 will conclude this paper with a discussion and outlook.

2   Related Work

One typical scenario in mobile learning is the support of field trips. Here, two major categories can be distinguished: The first category focuses on the activities in the field, e.g. by providing a learning application that would run on the mobile device. Chen and Kao's bird watching scenario [6] constitutes a typical example in this category. Here, learners use a handheld device (PDA), which is wirelessly connected to a bird-database and provides a learning application to support an outdoor bird-watching activity. In the second category, learners typically collect data (text, images, sensor data, etc.) in the field for further use in subsequent (classroom) activities. The LEMONADE system [7] is an example for an approach that supports field trip and classroom activities in these kind of scenarios. The approach presented in this paper relates more to the second category, as it can be applied in scenarios where data are collected in the field and visualized and post-processed in the classroom.

Another kind of learning environments utilizes mobile device to realize classroom response scenarios or supports classroom discussion or brainstorming sessions. Classroom response systems (CRS) are designed to support scenarios like multiple-choice quizzes or answering teacher questions with short text messages [8]. Numerous Classroom Response Systems have been created over the past years, using various types of devices, ranging from infrared senders with a few buttons to the use of smartphones running tailored applications. Classroom discussion and brainstorming support, as described e.g. in [9] is intended to enrich face-to-face discussions with features like group support, anonymous contributions, re-use and comparisons, automatic visualization, activity analysis and moderation support. MoCo can support classroom response and classroom discussion activities already now – future features, especially enhanced visualization function, may support advanced features of these scenarios.

Learning scenarios that include existing, commonly used software, including native applications for mobile devices can be found as well, e.g. for the use of Facebook [10] or Twitter [11] in educational settings. However, the authors are unaware of approaches that allow the combination of multiple input channels, including the use of third-party software as proposed here.

3   Design and Implementation

The technical infrastructure that was implemented to support different input dimensions is outlined in Fig. 1. The major task performed by this infrastructure is to

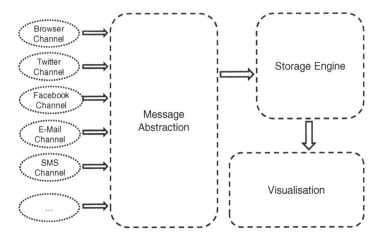

Fig. 1. Architecture to support multichannel input

provide a certain abstraction for the messages received through the different input channels, store these messages and to later-on allow a flexible message visualization to be used in (usually classroom based) learning units.

The following sections provide a more detailed description of each of the building blocks of the architecture as well as (an outline) of existing and planned input channels.

3.1   Message Abstraction

The central idea behind the development of the presented approach was the idea to make messages from different input channels available in a single architecture.

Messages from different input channels typically vary in certain features, e.g. their length, the availability of referring to other messages, additional information like GPS location etc. These variations call for a message abstraction to create convenient storage and retrieval mechanisms and to be able to generate a common visualization.

Still, all the messages from the different input channels have something in common, i.e. they provide specific, in most cases text-only, content. But already with respect to the content, the messages may differ dramatically: usually the size of the content of messages varies from channel to channel. E.g. a Twitter input channel usually consists of messages up to a length of 140 characters while short messages received by a mobile phone regularly have at least 160 characters and, furthermore, messages send e.g. via Facebook are not at all limited in their size. Still, different sizes of the messages may result from the user's choice of the input channel. This has to be considered in any review or evaluation of the different inputs, since therefore the length of a Twitter message has to be considered shorter (by design) than a message sent e.g. via Facebook.

In addition to the content itself, messages usually provide a certain set of data that is independent from the specific input channel. Supplementary, there are further metadata entries that are connected to the respective input channel that was used. In the presented implementation, the following set of basic data (including the content) was chosen to be used during message abstraction:

— MD5 hash: Used as a unique identifier. The hash value is based on the other basic metadata entries.
— Author: The person in charge of the publication of the content of the message.
— Content: The content of the message. If this is not text (i.e., a string), the binary content will be wrapped through Base64 encoding.
— Source type: The input channel via which the message was received.
— Time: A timestamp that indicates when the message was sent.
— Tags: The list of keywords/terms given to the message by the author.

A simple set of metadata that is used for specific channels could consist of entries like the following:

— inReplyTo: indicates whether the message was a reply to another message.
— (if available) Geo Location: longitude/latitude of author when sending a contribution (some Twitter or Facebook clients make this available).
— (if available) Language: language of message.

However, we do not find a predefined set of metadata that matches all of the foreseen input channels, e.g. the author of a message sent via the SMS service of a mobile phone will usually be represented by a phone number. In case of a message sent e.g. via Facebook, the user might be directly identifiable by his Facebook username.

3.2   Storage Engine and Backend Implementation

In order to increase the flexibility of message storing for the messages received through the different input channels, an abstraction layer was implemented that is responsible for storing messages. By this, we increased the ability of using different storing techniques, beside usual relational databases, e.g. XML files, Cloud Computing based storage or any other storage facility. Basically, this layer was implemented using the Builder Design Pattern [12].

Following the proposed infrastructure shown in Fig. 1, we have implemented a first prototype of the system. Even though there are currently only a few channels available, the backend was designed in such a way that it can support various additional channels in future. For the message abstraction layer and the communication between the components, we use the SQLSpaces [13, 14] implementation of the blackboard paradigm [15]. This allows us to loosely couple the different agents that will collect the input data. Another advantage of using the blackboard as the basis for the backend is the opportunity to attach software components that analyze learners' contributions, e.g. to detect specific patterns, to discover duplicates or spam, or to enrich contributions with additional metadata. As a relational database serves as the basis for the SQLSpaces we use this combination to store the contributions.

3.3   Implementation of Different Input Channels

In future versions of the MoCo system, different input channels will be integrated to allow contributions from different sources. One of the central motivations of allowing different input channels is to provide users with the possibility of using off-the-shelf software that is already installed on their (mobile) devices for creating messages in the context of the respective scenarios.

As we already described, the messages received by different input channels usually vary, e.g. in the length of the content. Also, these different input channels provide different internal structures for sending messages, e.g. messages on Facebook can be sent directly to the feed of a user or may be posted to a Facebook group, while SMS messages sent from a mobile phone, are usually not contextualized at all. Therefore, decisions where these messages are posted to, how to identify messages of interest and how to contextualize messages have to be taken. These decisions will be explained implicitly in the following subsections.

Twitter Client

The implementation of a Twitter client that receives messages and forwards these messages to the Message Abstraction and Storage components can easily be realized by using the publicly available Twitter API[1] or existing implementations for e.g. Java developers[2]. In order to identify messages sent to the described infrastructure, we decided to use hashtags, which are already known from Twitter or the IRC network, to tag messages referring to a certain course, group or task. Hashtags are simply created by adding the prefix '#' to any term in the message. More than one hashtag is possible in a message, which increases the flexibility of this approach and also allows for hierarchical tagging as shown in the following example messages:

"#seminar2012 #groupA I just understood queuing theory when standing at the checkout at the discounter!"

This message would relate to the terms (or keywords) 'seminar2012' and 'groupA'. We decided to include hashtags as known from Twitter as part of the messages' metadata processed in the message abstraction component. The approach of using hashtags does also work for other input channels and not only for the Twitter approach.

Facebook Client

A first implementation for an input channel allows retrieving messages that are posted to a certain Facebook group. Here, we used a publicly available Facebook API[3]. In order to get the necessary authentication for accessing a specific Facebook group, a

---

[1] Details about creating applications that integrate Twitter communication can be found at https://dev.twitter.com/ (last accessed April 23rd 2012).

[2] Twitter4J is a Java library to create Java applications that integrate Twitter services, see http://twitter4j.org (last accessed April 23rd 2012).

[3] Details about the used Facebook API can be found at http://www.restfb.com (last accessed April 23rd 2012).

Facebook application needs to be registered with Facebook[4]. The messages themselves are usually posted directly to the standard feed of the group. Identification of messages of interest is again implemented by the help of the formerly described hashtag approach. Once a certain message is received by the Facebook channel, this message is automatically forwarded to the already described abstraction of the storage engine. Here, the message is stored persistently.

SMS Client

Beside the possibility to integrate a costly SMS gateway, we decided to get access to SMS messages by deploying a simple Android based mobile phone. For this phone we developed a low footprint application that regularly checks for updates of the short message inbox of a mobile phone. In doing this, we implemented a cheap, flexible and easy to use SMS gateway that allows retrieving SMS sent to a conventional phone. Received messages can again be classified by the same tag that is already used to identify Twitter and Facebook message hashtags. Once such a message of interest is received, the message is passed to the storage engine in order to store the message persistently.

Web Client

In addition to the clients mentioned above, we implemented a web-based contribution channel. This web-channel allows users to store contributions without registering to Twitter or Facebook. It is implemented using JSP and JavaScript with the focus on ease-of-use and on compatibility for mobile devices like Apple iPhone or smartphones based on the Android OS. Fig. 2 shows screenshots from the web contribution client on different devices.

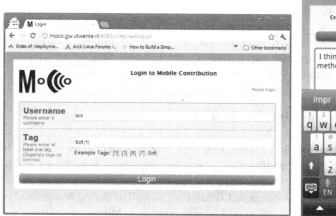

Fig. 2. Web contribution interface in a desktop client browser (left) and on an Android-based mobile device (right)

---

[4] Details about accessing information on Facebook can be found at
http://developers.facebook.com (last accessed April 23rd 2012).

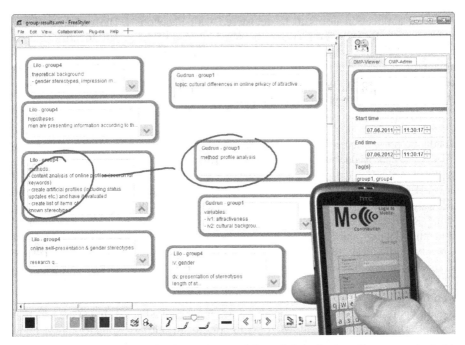

Fig. 3. FreeStyler as visualization component and the web contribution client on an Android-based smartphone

3.4   Visualization Frontend

To visualize the contributions, we use a plug-in for the drawing and modeling tool FreeStyler [16]. FreeStyler allows displaying the different contributions and ordering them according to the tags that the user specified for the contribution. Fig. 3 shows a screenshot of the FreeStyler environment with a number of sample contributions. Furthermore, the system enables the teacher to display only filtered contributions based on tags, time or author.

4   Experiences and First Evaluation

This section concentrates on first experiences with MoCo in university contexts. In [17] we described potential applications scenarios and their benefits to teachers/lecturers and students alike. As outlined in section 1, the tool can be used in and outside classrooms (spatial and situational flexibility) as well as during lessons and in students' spare time (formal vs. informal learning contexts). Moreover, learners can generate contributions individually or in groups in face-to-face settings, e.g. for immediate use in the seminar contexts, or use the tool in self-learning phases or for homework preparation where they might be distributed spatially and have different preferences to approach tasks. Thereby the tool makes it possible to collect, organize and work with the contributions also in a temporally flexible way (synchronous and asynchronous usage).

Two scenarios that were addressed as potential application situations in [17], namely the One-Minute Paper (OMP) and the support of self-learning phases, have been tested, and first experiences and evaluation results are outlined in the following. Due to the prototypical nature of the system's implementation at the time of evaluation, only the web-based contribution channel was used by the students via their mobile phones (if capable of showing web pages) or notebooks.

4.1   One-Minute Paper – Experiences with Exemplary Scenarios

OMP is a time-saving method for feedback collection that can be applied for groups of learners of all sizes and flexibly used at different stages of a lesson, e.g. to ensure that nothing remained unclear or to assess the learners' level of knowledge regarding a specific topic. Usually students' are given a piece of paper and about one to five minutes to note down their answers (anonymously). Contributions/Answers are collected afterwards [4, 5, 18]. Spoken from our own experience, learners are generally very enthusiastic about the idea of being asked for their opinion and having the chance to let the lecturer know about open questions (without disclosing them openly in class). However, application of this method needs at least some preparation beforehand (e.g. printing) and post-processing, e.g. to evaluate problems, collect open aspects, get an overview etc.

The electronic version collects students' contributions without preparation, does not require paper and pen, does not produce delays as everyone can make their contributions at the same time and can be easily managed afterwards with the Freestyler plug-in that allows filtering for specific words/questions numbers.

Since we were interested in potential differences in the number of words people would contribute (all in all), depending on whether the paper-pencil version was used or the MoCo version, we used the OMP in three different courses at two German universities. At the end of the lesson, students were instructed how to use MoCo and informed about the intention of the OMP. People were asked to answer three questions: (1) what was new to them (2) what they already knew and (3) what remained unclear. Those equipped with a device that could access the internet (e.g. own mobile phone or notebook) were asked to enter their answers via this device. The rest was asked to note their answers on the paper and hand it in anonymously.

If the analysis revealed a significant difference in the number of words students use, that way that people produce less words in the electronic variant of the OMP method, it might be considered as providing less information and insights compared to the pen and paper version.

Results from Applying the OMP Method

From the 20 students attending a Master's communication studies course (seminar), eight made their contributions via MoCo while 12 handed in the OMP on a piece of prepared paper with the three questions listed above. In the paper and pencil version people wrote a total number of 65 words for question one, 46 words for question two and two words for question number three (10 did not write anything). In the MoCo (technologically supported) version people altogether wrote 66 words for the first question, 29 words for question 2 and 40 words to answer question 3.

Besides the communication studies seminar, MoCo was used in two lectures for information science students. In the cryptography lecture, eight participants handed in their answers via MoCo while the other half used paper and pencil. Finally, six people attended the software lecture; three using MoCo for their OMP, three using paper and pen.

People answered the same three questions with the following word count for the paper and pencil version of the OMP: 41 (question 1), 19 (question 2), 18 (question 3) (for standard deviations please refer to Table 1). In the MoCo condition people answered a total number of 56 words with regard to the first question, 52 for question two and 36 words were counted for questions three. The second group of information science student wrote a number of 20 words to question one, 11 for question two and no word for questions three in the paper version, while 27 words (questions 1), 32 words (questions 2) and 16 words (questions 3) were counted in the MoCo OMP.

All in all, more than once, in all groups and conditions, people who obviously did not have any open questions (see word counts for question three) wrote „nothing" or drew a dash line to indicate that they did not want to say anything. As compared to the paper version, in the MoCo condition the total number of written words answered with regard to the questions was always higher, with one exception being question two in the paper condition (communication seminar) (please refer to Table 1 for details). While these results have certainly to be treated with care and do not allow a general conclusion of clear advantages or benefits of the MoCo supported OMP, it may be cautiously concluded that the data may be consistently interpreted in that way that MoCo does at least not elicit less contributions. However, this pattern needs some further empirical backup.

Table 1. Word count across conditions and questions in One-Minute Paper scenario

	Communication Seminar		Cryptography Lecture		Software Lecture	
	Paper	MoCo	Paper	MoCo	Paper	MoCo
Question 1	Total: 65 $M = 5.42$ $SD = 2.47$	Total: 66 $M = 9.43$ $SD = 5.50$	Total: 41 $M = 5.125$ $SD = 3.87$	Total: 56 $M = 7$ $SD = 3.21$	Total: 20 $M = 6.67$ $SD = 0.58$	Total: 27 $M = 9$ $SD = 5.57$
Question 2	Total: 49 $M = 5.11$ $SD = 3.33$	Total: 29 $M = 4.83$ $SD = 3.31$	Total: 19 $M = 2.38$ $SD = 1.77$	Total: 52 $M = 6.5$ $SD = 2.67$	Total: 11 $M = 3.67$ $SD = 2.08$	Total: 32 $M = 10.67$ $SD = 7.02$
Question 3	Total: 2 $M = 1$ $SD = 0$	Total: 40 $M = 13.33$ $SD = 3.06$	Total: 18 $M = 3.6$ $SD = 5.27$	Total: 36 $M = 5.14$ $SD = 4.09$	Total: 0	Total: 16 $M = 5.33$ $SD = 1.15$

4.2   Long Term Usage – "Take Home" Task.

Besides using MoCo to support the OMP scenario, we used and evaluated the system in the context of a Master's course on communications studies at the University of Duisburg-Essen to support phases of self-learning outside lectures. As outlined in [17], it is important for students to be able to integrate and "retrieve" prior knowledge with newly acquired (and the other way round) and to relate to situations in their everyday life. Students were asked to complete homework in a self-learning phase by using MoCo. The specific intention of the homework was to re-activate students' individual knowledge about "communication" in order to see what they remembered from the introductory Bachelor's course. Moreover, this task was intentionally placed at the beginning of the semester to be able to use the contributions people would make for further seminar lessons. The questions varied between open-ended sentences and situations that should be illustrated (compare example sentences below) in order to trigger different fields of knowledge in the student. This kind of task, on the one hand, provides the opportunity to compare the number of words and the style of writing people apply (in note form, whole sentences, etc.) as compared to pen and paper situations. However, this was not the focus of the present analysis. On the other hand, this initial use of the system lets students establish a first impression which they would be able to externalize in a questionnaire (as described in the following sections). With this in mind, students were asked to work on 11 tasks in the upcoming week, e .g. (1) My personal associations with the word communication are…, (3) Without communication it would be…, (5) Communication is so important for human beings because…, (7) The internet can help …, (8/9) This phenomenon/events of cmc has impressed me especially positive/negative, (11) Please describe a communicative misunderstanding.

In the following session FreeStyler (see Fig. 3) was used to organize the contributions (beforehand, in order not to lose too much time). People were asked for their impression on the task and about the benefits and drawbacks they felt in this regard. Afterwards they were asked to complete an online evaluation form that is described in the following.

Contents and Procedure

The online evaluation form was setup via ofB[5], a free and reliable tool for online studies. A link guided participants to the respective starting page of the online evaluation where they were informed about the aim of the evaluation, who is conducting the research, and that their data is used for scientific purposes only and treated anonymously.

Besides several demographic facts, e.g. their educational background, age and sex, we used items from the Technology Acceptance Model [19, 20] to evaluate Perceived Ease of Use and Perceived Usefulness, as acceptance of the system is a necessary requirement for widespread use.

Besides, we were especially interested in students' equipment with wireless internet access enabled mobile phones and notebooks and the situations in which they had

---

[5] For more details, visit `https://www.soscisurvey.de` (last accessed April 23rd 2012).

authored their contributions the week before. What was especially interesting to us as it informs the design of the tool, were their ideas about shortcomings of the software and how they could be addressed. Furthermore, they were asked to report their problems in using the system and to explain how they envision other application scenarios and improvements of the system's features.

4.3   Results

Participants: Demographics and Equipment

The group consisted of 14 participants 12 of which were female and two male. The mean age is M = 25.07 (SD = 2.70). Six people indicated to have a mobile phone that enabled them to access the internet, six had a mobile phone without internet connection and two said to have no mobile phone. 13 participants hold a university degree, one person indicated to have a university entrance qualification (German Abitur).

TAM: Perceived Usefulness and Perceived Ease of Use

In order to assess peoples' opinion about the system, we used the Perceived Usefulness and Perceived Ease of Use items. Perceived Ease of Use and Perceived Usefulness are measured by seven statements about the tool, e.g. "MoCo is easy controllable and behaves as expected." on a 5-point Likert scale ranging from "0 = do not agree at all" to "4 = fully agree". Cronbach's alpha was calculated to determine the internal consistency of the items and showed to be good for the seven Perceived Ease of Use items α = .824 and rather mediocre for the seven Perceived Usefulness items α = .617.

Items of both scales were summed up to calculate the average values. All in all, both Perceived Usefulness reached M = 2.88 (SD = .26) and Perceived Ease of Use M = 2.95 (SD = .41) reached mediocre average values with a tendency towards the positive side. So this leaves some room for improvements and more positive evaluations.

Devices Used, Conditions and Situations of Use

With regard to the devices people used to make their contributions, 13 people indicated to have used a PC/laptop, three had used mobile phones, no one indicated to have used an iPad/tablet (multiple answers possible). In addition, we asked people to indicate on a 5-point Likert scale (1=never, 5=always) how often specific situations applied when they were using MoCo. Table 2 gives an overview about the answers. The pattern that can found in the data is that people tend to have been alone when producing their contributions rather than being surrounded by others. Also, people tend to have been at home while making contributions rather than on the train/in the bus, on the campus or elsewhere (parties, restaurants, etc.).

Contribution Time

Since we were interested if people would prefer a specific time of the day for the preparation of the MoCo contributions, we asked them for the daytime at which most of their contributions were made. People could choose between: in the morning (N=2), at lunchtime (N=4), in the evening (N=5) and at all times/not at a specific time (N=3). Results show that people seem to have varying preferences. The same seems to be true for the time that passes by before people write their entries. People could choose from a number of statements and were asked to select the one that best fitted their behavior: I always sent my contribution immediately when I had an idea fitting the task (N=2), I made up my mind about the task at one specific moment and directly sent in all my contributions (N=9), One way or the other (N=3).

Table 2. Conditions of MoCo use

Item	Frequency of participants choosing the different options			M	SD
I was alone.	1 (N=1) 2 (N=1) 4 (N=2) 5 (N=9)		3 (N=1)	4.21	1.31
I was surrounded by others.	1 (N=5) 2 (N=1) 4 (N=3) 5 (N=2)		3 (N=3)	2.71	1.54
I was on the train /in the bus.	1 (N=1) 2 (N=0) 4 (N=0) 5 (N=13)	3 (N=0)		1.29	1.07
I was on campus (lecture hall, library, etc.).	1 (N=7) 2 (N=1) 4 (N=2) 5 (N=3)		3 (N=1)	2.50	1.74
I was at home.	1 (N=1) 2 (N=0) 4 (N=3) 5 (N=9)	3 (N=1)		4.36	1.15
I was on the way (shopping, etc.).	1 (N=13) 2 (N=0) 4 (N=0) 5 (N=1)	3 (N=0)		1.29	1.07
I was at a party, at the cinema, in a restaurant, etc.	1 (N=13) 2 (N=0) 4 (N=0) 5 (N=1)	3 (N=0)		1.29	1.07

5   Discussion and Outlook

In this paper, we described a flexible mobile learning environment that allows students to create and send contributions via multiple input channels and outlined first experiences in using the software in and outside classrooms. We argued that to reach large numbers of learners and teachers, a successful system needs to address existing and widely used software (such as Facebook or Twitter) and widespread hardware (such as iPhones, Android-based smartphones or tablets), and that it needs to get on with minimal efforts regarding configuration and maintenance. The proposed architecture and implementation allows the realization of different kinds of scenarios without configuration or customization, as the MoCo environment allows working synchronously and asynchronously, supports working in groups or individually (e.g. by using group tags), and can be used in classroom situations as well as "in the field".

A first working prototype has been presented that is able to collect students' contribution with the help of an SQLSpaces blackboard architecture and that visualizes contribution with the help of the FreeStyler application.

The OMP scenario suggested in previous literature was used in an (authentic) academic seminar context, as well as in two lectures with the aim of assessing students' spontaneous reactions to the method and quantitatively comparing the number of words written depending on whether people were using MoCo or paper and pencil. Reactions were generally positive and word count comparisons indicate (at least for the students of the investigated sample) that the MoCo supported OMP does regularly result in a higher average number of words compared to the paper and pencil version. While it is interesting that this result applied independent from the content of the course and number of people contributing, this result has to be treated with care and validated by larger samples to ensure that this was not e.g. a novelty effect going along with an increased motivation to use mobile phones and computers in class or caused by the specific composition of students who are attracted by technology and new media. A challenge that also became obvious from this first trial is that many students still do not have a mobile phone with internet access or carry their notebook with them at the university. However, we expect that this will change in near future as part of a general trend of an increased penetration of mobile devices with internet access especially among young people.

With regard to the take home task and the succeeding evaluation of Perceived Ease of Use and Perceived Usefulness of the system and the specific usage characteristics, we gained some useful insights. Especially people's description of problems with the system and ideas for its enhancement are valuable for the next steps in development. Their implementation will hopefully lead to higher attributions of usefulness and ease of use. Also, people have made interesting suggestions for further scenarios. All in all, further scenarios should be tested in small and large groups to see where the limits of MoCo are, what kind of tasks with how many people can be reasonably supported without e.g. producing too many contributions for a manageable post-processing, ad-hoc use. First experiences with the evaluation questionnaire have also shown that with larger groups it may be interesting to see whether there are specific usage patterns and whether these vary with people's gender, the preferred contribution time and location or specific personality traits. Future evaluations could also include motivational aspects and further investigations of students and teachers actual needs. Moreover, as already suggested earlier, valuable conclusions may be drawn from qualitative analyses of students' contributions' contents.

On a technical level, our next steps will include the implementation of additional input channels or the improvement of existing ones, taking especially into account the feedback of our first generation of users. We plan to use cloud services to increase the reliability and scalability of our components. For the visualization, we foresee a web-based (JavaScript) implementation, rather than the current Java application, to minimize installation efforts and to maximize compatibility.

References

1. Gerpott, T.J.: Attribute perceptions as factors explaining Mobile Internet acceptance of cellular customers in Germany – An empirical study comparing actual and potential adopters with distinct categories of access appliances. Expert Systems with Applications 38, 2148–2162 (2011)
2. Kukulska-Hulme, A., Sharples, M., Milrad, M., Arnedillo-Sánchez, I., Vavoula, G.: The genesis and development of mobile learning in Europe. In: Parsons, D. (ed.) Combining E-Learning and M-Learning: New Applications of Blended Educational Resources, pp. 151–177. Information Science Reference (an imprint of IGI Global), Hershey (2011)

3. Chen, Y.S., Kao, T.C., Yu, G.J., Sheu, J.P.: A Mobile Butterfly-Watching Learning System for Supporting Independent Learning. In: Proceedings of 2nd IEEE International Workshop on Wireless and Mobile Technologies in Education, pp. 11–18. IEEE Computer Society (2004)

4. Chizmar, J.F., Ostrosky, A.L.: The One-Minute Paper: Some Empirical Findings. Journal of Economic Education 29, 3–10 (1998)

5. Stead, D.R.: A Review of the One-Minute Paper. Active Learning in Higher Education 6, 118–131 (2005)

6. Chen, Y.S., Kao, T.C., Sheu, J.P.: A mobile learning system for scaffolding bird watching learning. J. Comput. Assist. Lear. 19, 347–359 (2003)

7. Giemza, A., Bollen, L., Hoppe, H.U.: LEMONADE: field-trip authoring and classroom reporting for integrated mobile learning scenarios with intelligent agent support. International Journal of Mobile Learning and Organisation 5, 96–114 (2011)

8. Roschelle, J.: Unlocking the learning value of wireless mobile devices. J. Comput. Assist. Lear. 19, 260–272 (2003)

9. Bollen, L., Juarez, G., Westermann, M., Hoppe, H.U.: PDAs as Input Devices in Brainstorming and Creative Discussions. In: Proceedings of International Workshop on Wireless, Mobile and Ubiquitous Technologies in Education (WMUTE 2006), pp. 137–141. IEEE Computer Society (2006)

10. Yang, Y., Wang, Q., Woo, H.L., Quek, C.L.: Using Facebook for teaching and learning: a review of the literature. International Journal of Continuing Engineering Education and Life-Long Learning 21, 72 (2011)

11. Despotovic-Zrakic, M.S., Labus, A.B., Milic, A.R.: Fostering enginering e-learning courses with social network services, 122–125 (2011)

12. Gamma, E., Helm, R., Johnson, R., Vlissides, J.: Design Patterns. Elements of Reusable Object-Oriented Software. Addison-Wesley (1995)

13. Weinbrenner, S., Giemza, A., Hoppe, H.U.: Engineering Heterogeneous Distributed Learning Environments Using Tuple Spaces as an Architectural Platform. In: Proceedings of Seventh IEEE International Conference on Advanced Learning Technologies (ICALT 2007), pp. 434–436 (2007)

14. Bollen, L., Giemza, A., Hoppe, H.U.: Flexible Analysis of User Actions in Heterogeneous Distributed Learning Environments. In: Dillenbourg, P., Specht, M. (eds.) EC-TEL 2008. LNCS, vol. 5192, pp. 62–73. Springer, Heidelberg (2008)

15. Buschmann, F., Meunier, R., Rohnert, H., Sommerlad, P., Stal, M.: Pattern-Oriented Software Architecture: A System of Patterns. John Wiley & Sons Ltd., West Sussex (1996)

16. Hoppe, H.U., Gaßner, K.: Integrating Collaborative Concept Mapping Tools with Group Memory and Retrieval Functions. In: Proceedings of Proceedings of the International Conference on Computer Supported Collaborative Learning (CSCL 2002), pp. 716–725. International Society of the Learning Sciences (2002)

17. Bollen, L., Jansen, M., Eimler, S.C.: Towards a Multichannel Input Dimension in Learning Scenarios with Mobile Devices. In: Proceedings of 7th IEEE International Conference on Wireless, Mobile and Ubiquitous Technologies in Education (WMUTW 2012), pp. 311–315. IEEE Computer Society (2012)

18. Light, G., Cox, R.: Learning and Teaching in Higher Education: The Reflective Professional. Paul Chapman, London (2001)

19. Davis, F.D.: Perceived Usefulness, Perceived Ease of Use, and User Acceptance of Information Technology. MIS Quarterly 13, 319–340 (1989)

20. Davis, F.D., Bagozzi, R.P., Warshaw, P.R.: User Acceptance of Computer Technology: A Comparison of Two Theoretical Models. Management Science 35, 982–1003 (1989)

Software Requirements to Support QoS in Collaborative M-Learning Activities

Didac Gil de La Iglesia, Marcelo Milrad, and Jesper Andersson

DFM, Linnaeus University, Växjö, Sweden
{didac.gil-de-la-iglesia,marcelo.milrad,jesper.andersson}@lnu.se

Abstract. The use of collaborative activities in education has proven to be an effective way to enhance students' learning outcomes by increasing their engagement and motivating discussions on the learning topics under exploration. In the field of Technology Enhanced Learning (TEL), the use of information and communication technologies has been extensively studied to provide alternative methods to support collaborative learning activities, combining different applications and tools. Mobile learning, a subset of TEL, has become a prominent area of research as it offers promising tools to enhance students' collaboration and it provides alternative views for teaching and learning subject matter in relevant and authentic scenarios. While many studies have focused on the pedagogical opportunities provided by mobile technologies, fewer are the efforts looking at technological related aspects. Hardware and software issues in this field still remain as challenges that require a deeper level of study and analysis. This paper presents and discusses the findings of a deep analysis based on the outcomes of three mobile collaborative learning activities and their requirements. These results have helped us to identify a number of arising challenges that need to be addressed in order to warranty Quality of Service (QoS) in these collaborative M-learning activities. Moreover, the paper offers a view on current practices in M-learning activities, which evidences the lack of research addressing software engineering aspects in mobile collaborative learning.

1   Introduction

The use of Information and Communication Technologies (ICT) in education is not new but progressing at a pace that is driven by advances in other areas, as expressed in the literature [8]. Recently, ICT has expanded to encompass a more complex working environment than the traditional classroom setting. Outside the classrooms, and located in more authentic environments, mobile technologies allow students to explore alternative ways to understand subject matter related issues. Mobile computing technologies allow users mobility while carrying such a computational device. Consequently, the use of mobile computing technologies in the field of education, known as M-Learning [10], is progressing significantly. This is demonstrated by the number of relevant efforts have been carried out related to the use of mobile technologies in education focusing on pedagogical aspects

V. Herskovic et al. (Eds.): CRIWG 2012, LNCS 7493, pp. 176–183, 2012.

[8, 18, 19, 11], and on the different uses of new mobile tools and applications [2, 16]. However, the number of studies focusing on these efforts are not aligned with the ones focusing on the technological issues related to M-Learning.

Indeed, M-Learning activities offer new possibilities to education, such as new ways to support collaboration among the students. However, the adoption of mobile computing technologies may also imply the addition of some new problems and challenges. The dynamic environments where M-Learning activities take place entails uncertainties due to lack of knowledge during development about the events and environment [7] that may affect technological services and threatens collaboration. Addressing these uncertainties and the derived risks is a crucial task that needs to be studied to guarantee that activity functionalities are achieved, such as collaboration, in order to obtain the desired goals.

Collaboration becomes a cornerstone [3] in the learning process, motivating students to *"take on roles, contribute ideas, critique each other's work, and together solve aspects of larger problems, all to good effect"* [13]. Moreover, it incentives learners to argument for their ideas, thus providing solid foundations and aligning them with previous acquired skills and knowledge. Thus, TEL applications should consider and support collaboration in order to bring its benefits to everyday educational practices. From a technological point of view, there are two separate components that need to be shared to support collaboration in mobile learning activities, namely the activity itself and the resources allocated to support it. Resources can be hardware (camera, GPS, memory, processor unit, etc.), software (media files, activity tasks and outcomes, messages, etc.) or communication channels (access to the Internet, access to internal servers, etc.) to support for collaboration [15].

However, as mentioned before, uncertainties are present in software development, threatening, for example, the sharing of enumerated resources. Chances are even higher in M-Learning activities, as several variables come into action: user's mobility, fragility of devices, use of batteries, low software and hardware reliability. Therefore, it becomes a challenge to guarantee the availability of the required activity resources. Such guarantees are described in terms of QoS levels, which the system is expected to provide. In the field of M-Learning, there have not been enough efforts to create and develop software solutions focused on providing some guarantees to support these properties.

This study elaborates on the software engineering aspects that are necessary to support the level of collaboration demanded by M-Learning activities. Therefore, one of our main goals is to identify a set of requirements necessary in order to guarantee that the required collaboration can be offered. Thereafter, the study presents possible strategies to address the mentioned uncertainties and strength the desired QoS guarantees. In order to address this objective, the paper describes and discusses the characterization of different scenarios that involve mobile technologies in collaborative outdoor activities. From this analysis, we bring up a set of identified uncertainties, present in M-Learning activities, that may put collaboration at risk. Finally, the study complements these results with two extensive systematic literature surveys in order to present possible mitigation approaches to guarantee QoS in M-Learning applications.

2   Motivation

A collaborative mobile learning activity includes a set of functional objectives to be accomplished that may vary depending on the content and pedagogical approach undertaken. However, there are several uncertainties that cause risks that may affect the achievement of these learning activities. The functional objectives, including collaboration, defined by teachers and students require the support of non-functional requirements to provide some guarantees. It is of interest to provide a certain level of QoS in the M-Learning activities aimed at collaboration, which could be understood as Service Level Agreements between relevant stakeholders involved in the activity. In indoor traditional activities (using books and blackboards), and in front of unexpected problems, the activity flow can be assured to some extent by adapting the formulation of the activity, changing some parameters on the blackboard and applying different pedagogical strategies. In indoor TEL activities, the complexity to maintaing the learning activity increases accordingly of potential points of failure, but it is still feasible to address. However, adapting outdoors mobile collaborative activities to guarantee its correctness becomes an arduous task, as it would require new system reconfiguration and possibly redesign of the application and redeployment on devices. Incidents during outdoor collaborative M-learning activities can put at risk the execution of the lessons and, consequently, the expected pedagogical outcomes. Therefore it is relevant to determine the characteristics that define the collaborative M-Learning activities and to analyze to which extend these characteristics have been taken into consideration in the design, development and deployment of the current M-Learning software systems.

In GEM (GEometry Mobile), a collaborative M-learning project [5], learners worked in teams measuring distances to perform geometrical calculations (lengths, areas and volumes) using mobile devices and customized mobile applications. The experience and knowledge gained during the activity evidenced that some technical factors had a major impact in the flow of the activity and in some cases did put it in a state that threatened the successful completion of the learning tasks. One was the risk of a student closing the M-Learning application while the activity was still running. Generally, learners in our M-Learning activities are characterized by students in between the ages of 8–16 year old. They are prone to perform unexpected and undesired actions on the device, such as opening games during the learning activity, which could imply closing the M-learning application. This event influenced not only the student carrying the affected mobile device but also the collaboration inside the group. The physical environment and its conditions contributed also with uncertainties that had an effect on the activity as well. As an example, meteorological conditions affected the accuracy of GPS devices, which in turn provoked a degradation of the functionalities needed in GEM. The existence of uncertainties and risks is confirmed by similar projects reported in the literature. For instance, The Manhattan Story Mashup [21], a collaborative story telling activity where the participants shared pictures taken from their mobile devices to be presented on public displays, experienced server problems that led to 30 minutes' interruption in which all participants were affected.

In the field of M-Learning, several efforts have been conducted to support collaboration. However, most of these solutions have looked at the field from an idealistic perspective [21, 14, 12, 8, 20, 9, 17, 15]. In most of the proposed solutions, they have considered the application to be used under optimal conditions, which is far from reality. In the following section, we present the results emerging from GEM and used them to identify software requirements and challenges that are required to guarantee QoS.

3   Software Requirements and Challenges

The analysis presented in this section provides a view of functional and non-functional requirements that were required for the creation of three collaborative M-Learning activities in the field of mathematics, and derives a set of challenges to be discussed in the next section. This requirement elicitation covers a number of M-Learning activities where the mobile device is used as a supporting tool for the students to carry and utilize during the entire learning activity.

In our previous efforts [6], an study of three iterations of the GEM project was presented. These three iterations are based on a life-cycle incremental development, where new requirements were included based on deficiencies and activity evolutions that needed to be addressed. Derived from these efforts, we specified a set of final requirements that are requirements (R) for GEM (Table 1). A subset of the requirements listed in this table is common in M-Learning, such as context acquisition related requirements. However, we have identified requirements that are crucial for the M-Learning activities, such as group collaboration and aspects related to QoS.

Table 1. Requirements identified in the GEM project

Req. ID	Name	Category
R1	Location acquisition	Context Acquisition
R2	Group collaboration	Organization
R3	Remote access to GPS coordinates	Service Sharing
R4	Remote access to a display on a mobile device	Service Sharing
R5	Remote access to media files	Service Sharing
R6	Multiple behavior device	Organization
R7	Log availability	QoS
R8	Activity Script Availability	QoS
R9	System tolerant to changes in the environment	QoS
R10	System tolerant to changes in the subsystem	QoS
R11	System tolerant to changes in the activity	QoS

The collaborative aims, the lack of functionalities in the mobile devices and the dynamism of the activity environment provided some evidences that there are aspects that require a high level of attention in the field. These can be classified into the following three categories: Service Sharing, Organization Management, Resilience (as a QoS) to availability issues.

To guarantee collaboration in M-Learning activities, certain QoS properties must be addressed in our software engineering solutions. As argued in [4], Availability, Reliability and Performance are concerns that are present in collaboration, precisely to address the desired QoS properties. For example, remote access to GPS coordinates implies the availability of the GPS resource as well as the reliability of the information the GPS service provides. Changes in the activity at runtime can affect collaboration, and the list of resources to be shared. Therefore, an adaptation mechanism needs to be provided to fix the availability and reliability of the shared resources, so the collaboration performance is not substantially affected. Failures in the system, due to devices disconnecting, resources becoming unavailable or environment conditions that deteriorate the services become challenges that deserve more attention.

The kind of problems described above are not unique for collaborative M-Learning activities, but shared amongst fields using distributed software solutions in dynamic environments. However, their impact in M-Learning is significant, as collaboration with mobile devices requires a distributed system [6] and the highly dynamic environment is implicit in them. Therefore, M-Learning becomes an interesting field to study how QoS properties can be guaranteed for distributed and dynamic systems. The complexity inherent in M-Learning activities becomes a challenge that requires the use of formal methods to analyze and study possible mechanisms for mitigating the potential risks. The following section presents and discusses possible strategies that can be implemented in oder to guarantee QoS for these kind of problems.

4   Possible Mitigation Strategies and Related Efforts

In front of the problems that may arise during the implementation of the system, we must find mitigations to address them. Modern systems can become too complex to be maintained manually. Moreover, if QoS wants to be achieved, response time to avoid risks or face undesired system states can be a critical factor to consider. Approaches to provide systems' adaptations autonomously and at runtime are known as Self-adaptive systems. Recently we have carried out an initial study [4] exploring which methods can be used for the design of SAS and why to use them. The work presents different mitigation mechanisms oriented to cover requirements in M-Learning activities, such as the presented in the previous section. Serviceability, design choice, modularity, generality, and upgradeability mechanisms can be considered as mitigation mechanisms to the potential risks that M-Learning activities may face. The presented work argues for the use of these mitigations and presents an initial self-adaptive design based on the use of the enumerated mitigations.

However, the success of a self-adaptive system design is conditioned to the effects that the environment may cause. Designing a system that adapts its behavior to address potential failures is an complex task, due to the number of variables that can affect the system and the potential combinations that could occur. Therefore, it is necessary to apply formal methods to evaluate the correctness of the system. By using formal methods, it is possible to *"rigorously specify and verify the behavior of self-adaptive systems"* [22].

In a recent effort [22][1], we have conducted a systematic survey including publications from 2000 until 2011 published in 16 well-recognized venues specialized in the field of Self-Adaptation to identify works using formal methods to study self-adaptive system properties. 75 studies match these specific research criteria, but the study identified that none of those had the Education field as an application domain. The first survey was complemented by analyzing studies performed in the field of TEL and M-Learning. A second literature survey [4][2], selected forty studies from 6 venues to determine the use of self-adaptive systems on the field of Mobile Computing and M-Learning from 2007 until 2011, considering the youth of the M-Learning field. In general, the studies consider goals as static elements that can be analyzed during the design phase and implemented in the development phase. However, goals cannot be static, but they must be modifiable, removable and able to be added at runtime [1], and our experience in GEM supports this statement. The activity flow and the activity content have been the most considered aspect in M-Learning to include self-adaptation mechanisms, to fit with the student's context, due to the pedagogical benefits it can bring. In the Mobile Computing field, there are studies that focus on self-adaptation oriented to technological aspects mainly adapting protocols and compression mechanisms to provide the proper QoS in hypermedia delivery. In these studies, self-adaptation mechanisms have been applied to provide resilience towards the connectivity along mobile devices by transmission route adaptation. However, as previously mentioned, we did not find studies in which the concept of resilience has been applied to assure functional requirements in relation to collaboration [4].

After analyzing a total of 22 venues, both from the field of Software Engineering and M-Learning, we have identified that self-adaptation has not been applied to cover functional requirements with regard to collaboration in M-Learning activities. Moreover, in the cases where self-adaptation has been considered, formal methods have not been applied to validate the desired properties of the system previous to execution, or to identify vulnerabilities that should be redesigned post-implementation.

5   Conclusions

Collaboration has been proven to be a beneficial feature to implement in educational activities. Moreover, the fast evolution of ICT related technologies and their adoption is providing promising tools for TEL and specifically in the field of M-Learning. The merge of these two lines of efforts is becoming noticeable, bringing collaborative activities in M-Learning, thus paving the road for a promising future regarding the development of collaborative M-Learning applications. It is, therefore, reasonable to consider that applications used in such activities should support all aspects related to collaboration if the expected pedagogical outcomes want to be guaranteed. However, there are multiple uncertainties that can risk

---

[1] Selection criteria and results can be found in [22].
[2] Selection criteria and results can be found in [4].

the achievement of such activities' goals, such as changes in the environment, device failures and modifications of the activity scripts at runtime. The dynamism always present in M-Learning activities, and even more prominent in outdoors activities, embraces risks that can affect the activities and provoke unexpected failures with respect to supporting the collaboration. These aspects demand systems that can become tolerant to changes during execution. This study suggests the use of self-adaptive mechanisms as a potential solution to resolve some of the risks that can be present in such activities and environments to provide certain QoS. To our best knowledge, previous studies in the field have not considered non-functional requirements, such as availability and reliability, to address the inherent risks that dynamicity of collaborative M-Learning activities imply. The results obtained from the analysis presented in this study show that there is a critical need to increase the consideration of self-adaptive mechanisms and, secondly, to apply formal methods to validate the correctness of the system.

Future efforts in relation to the work presented in this paper include the adoption of self-adaptation mechanisms in such mobile applications. Several are the self-adaptation efforts that have been presented in the software engineering field. Therefore, a study regarding the implications and benefits of self-adaptation is required to identify which mechanisms are suitable for our implementations and how can these be combined to provide resilience for our multiple functional requirements. An evaluation on the adoption of self-adaptation mechanisms in collaborative mobile learning applications will be a following stage in our efforts in order to validate the benefits of these solutions towards achieving our purposes. The field of TEL would benefit from such solutions as mobile applications will become more reliable in supporting collaboration for the implemented learning activities.

References

[1] Aseere, A.M., Millard, D.E., Gerding, E.H.: Ultra-Personalization and Decentralization: The Potential of Multi-Agent Systems in Personal and Informal Learning. In: Wolpers, M., Kirschner, P.A., Scheffel, M., Lindstaedt, S., Dimitrova, V. (eds.) EC-TEL 2010. LNCS, vol. 6383, pp. 30–45. Springer, Heidelberg (2010)
[2] Chan, T., et al.: One-to-One Technology-Enhanced Learning: an Opportunity for Global Research Collaboration. Research and Practice in Technology Enhanced Learning 1(1), 3–29 (2006)
[3] Dillenbourg, P.: Collaborative Learning: Cognitive and Computational Approaches. Advances in Learning and Instruction Series. Elsevier Science, Inc. (1999)
[4] Gil de la Iglesia, D.: Uncertainties in Mobile Learning applications: Software Architecture Challenges. Licentiate thesis, Linnaeus University (2012), http://lnu.diva-portal.org/smash/get/diva2:524874/FULLTEXT01
[5] Gil de la Iglesia, D., et al.: Enhancing Mobile Learning Activities by the Use of Mobile Virtual Devices – Some Design and Implementation Issues. In: Proceedings of INCoS 2010, pp. 137–144. IEEE Computer Society (November 2010)
[6] Gil de la Iglesia, D., et al.: Towards a Decentralized and Self-Adaptive System for M-Learning Applications. In: Proceedings of WMUTE 2012. IEEE Computer Society, Takamatsu (2012)

[7] Hastings, D., McManus, H.: A framework for understanding uncertainty and its mitigation and exploitation in complex systems. IEEE Engineering Management Review 34(3), 1–19 (2006)

[8] Herrington, J., et al.: Using mobile technologies to develop new ways of teaching and learning. In: New Technologies, New Pedagogies: Mobile Learning in Higher Education, pp. 1–14. University of Wollongong (2009)

[9] Herskovic, V., et al.: Modeling groupware for mobile collaborative work. In: Proc. of CSCWD 2009, pp. 384–389. IEEE Comp. Soc., Santiago (2009)

[10] Jones, V., Jo, J.: Ubiquitous learning environment: An adaptive teaching system using ubiquitous technology. In: Proceedings of ASCILITE 2004. Beyond the Comfort Zone, pp. 468–474. Australiasian Society for Computers in Learning in Tertiary Education (2004)

[11] Kukulska-Hulme, A., et al.: Innovation in Mobile Learning: A European Perspective BT. Inter. Journal of Mobile and Blended Learning 1(1), 13–35 (2009)

[12] Kurkovsky, S.: Multimodality in Mobile Computing and Mobile Devices: Methods for Adaptable Usability. IGI Global (2010)

[13] Looi, C.K., et al.: Collaborative activities enabled by GroupScribbles (GS): An exploratory study of learning effectiveness. Computers & Education 54(1), 14–26 (2010)

[14] Neyem, A., Ochoa, S.F., Pino, J.A., Franco, D.: An Architectural Pattern for Mobile Groupware Platforms. In: Meersman, R., Herrero, P., Dillon, T. (eds.) OTM 2009 Workshops. LNCS, vol. 5872, pp. 401–410. Springer, Heidelberg (2009)

[15] Neyem, A., et al.: A Patterns System to Coordinate Mobile Collaborative Applications. Group Decision and Negotiation 20(5), 563–592 (2011)

[16] Ogata, H., et al.: Computer supported ubiquitous learning environment for vocabulary learning. International Journal of Learning Technology 5(1), 5–24 (2010)

[17] Rogers, Y., et al.: Enhancing learning: a study of how mobile devices can facilitate sensemaking. Personal and Ubiquitous Computing 14(2), 111–124 (2010)

[18] Sharples, M., Taylor, J.: Towards a theory of mobile learning. In: Proc. of mLearn 2005 (2005)

[19] Sharples, M., et al.: Mobile Learning. In: Balacheff, N., et al. (eds.) Technology-Enhanced Learning, pp. 233–249. Springer, Netherlands (2009)

[20] Tarkoma, S.: Mobile Middleware: Architecture, Patterns and Practice. John Wiley & Sons Inc. (2009)

[21] Tuulos, V.H., Scheible, J., Nyholm, H.: Combining Web, Mobile Phones and Public Displays in Large-Scale: Manhattan Story Mashup. In: LaMarca, A., Langheinrich, M., Truong, K.N. (eds.) Pervasive 2007. LNCS, vol. 4480, pp. 37–54. Springer, Heidelberg (2007)

[22] Weyns, D., et al.: A Survey on Formal Methods in Self-Adaptive Systems. In: Proc. of FMSAS 2012 (2012)

Systems Integration Challenges for Supporting Cross Context Collaborative Pedagogical Scenarios

Dan Kohen-Vacs[1,2], Arianit Kurti[2], Marcelo Milrad[2], and Miky Ronen[1]

[1] Holon Institute of Technology, Israel
mrkohen@hit.ac.il
[2] CeLeKT, LinnaeusUniversity, Sweden

Abstract. This paper discusses the potential and challenges of integrating collaborative and mobile technologies in order to support a wide variety of learning activities across contexts. We present and illustrate two examples of such integrations aiming to expand the functionalities of an existing CSCL environment by introducing mobile technologies. Our goal is to enable the design and enactment of pedagogical scenarios that include asynchronous learning, outdoor collaborative activities and tasks performed in class using personal response systems. These examples are used to identify and analyze different challenges related to software systems integration issues. The outcome of these efforts is a proposed cross context systems integration model that can serve as the basis for future work that leads towards the integration of additional mobile applications designed and implemented to support novel collaborative learning scenarios.

Keywords: systems integration, pedagogical scripts, learning across contexts.

1   Introduction

Current developments in information and communication technologies (ICT) offer support for a variety of pedagogical activities performed across different spaces (indoors, outdoors and virtual) and involving various social settings (individual, small groups, class or community) [1]. Such learning processes described by a pedagogical scenario may be depicted within a pedagogical script [2]. These scripts may include interrelated phases that are based on each other and could be performed across different spaces encompassing diverse social settings. Furthermore, the script's phases may require various technologies that cope with the different challenges presented across its context. For example, home activities would be better performed asynchronously using stationary computers connected to the internet; classroom activities lead by the teacher as part of a lecture may be supported by personal response systems (PRS) [3] and outdoor activities including observation and data collection may use mobile GPS enabled devices [4]. Pedagogical cross-context scripts bring numerous challenges from a system integration perspective. This integration process involves several aspects like data exchange that provides essential data connectivity between standalone applications and workflow integration that deals with logic rules between the pedagogical script elements of different systems.

V. Herskovic et al. (Eds.): CRIWG 2012, LNCS 7493, pp. 184–191, 2012.
© Springer-Verlag Berlin Heidelberg 2012

This paper reports our on-going efforts for integration of different technological environments for supporting scripts that foster a collaborative learning process performed across contexts. The paper is organized as follows; the next section briefly describes the CeLS (Collaborative e-Learning Structures) platform for designing and enacting online collaboration using scripts. CeLS serves as a main catalyst platform for integrating the different mobile applications discussed in this paper. Specifically, it will present different examples of the abilities to design and enact integrated pedagogical scripts that include cross context, currently used by teachers and CSCL researchers. The following section describes and discusses the identification of a number of key issues and challenges related to the systems integration process. Finally, in the last section, concluding remarks are drawn and future steps are put forward towards addressing new integration challenges while implementing novel collaborative learning scenarios.

2   Fostering Systems Integration Design

Systems integration involved in cross context collaborative learning scripts should be profoundly grounded both, pedagogically and technologically [5]. In the coming examples described in this section, we illustrate how we used CeLS platform as the main catalyst for integrating a couple of mobile applications to support different cross context learning scripts. CeLS enables design, enactment and reuse of CSCL scripts [6], as well as it supports the reuse of data within the learning process [7].

A CeLS script is an XML based data description that defines the scenario's elements: the general framework of the activity, its phases, building blocks, special logic rules and finally the definitions of different elements' relationship across phases [6]. The script also comprises with the detailed social setting description for each one of the elements. A script expresses the depiction of the learning activity and its needed resources that could be enacted by CeLS or with other environment. The script is interpreted by the CeLS runtime engine that accordingly generates user-dedicated information and interactive interfaces that contain texts, websites, pictures or movie clips (and react to them later on). The interfaces may contain requests for learners' contribution according to a posed question.

Learner's contribution can be submitted in different formats like text and media. The contributions could also provide students with a mean for peers' evaluation, comments and feedbacks. The submitted data is stored within CeLS database and marked with a unique data identifiers built from the activity, phase and building block identifier manifested in the CeLS script. Another type of CeLS script element may originate in a form of request for interaction with artifacts that were previously contributed by other participants. This type of interaction could be accomplished using an interaction type of building block. This building block is marked by a unique identifier that specifies the elements it interacts with (from a previous phase) and information about the social settings associated with the interaction.

CeLS was originally designed to support asynchronous learning activities performed with stationary computers or laptops. However, as mentioned before, a script may also include notations that describe actions to be performed with external standalone applications. The attachment of the unique identifiers plays a key factor when integrating external environments with CeLS. It enables stateful representation by allowing CeLS to keep track of the generated artifacts initialization and processing state by setting its internal values (i.e. unique identifiers). This is very important for the reuse of the artifacts across different phases of the script using various technological devices and services as illustrated in the following examples. Fig. 1 illustrates the data exchange between CeLS and other standalone applications. Such data exchange is supported by XML based data format.

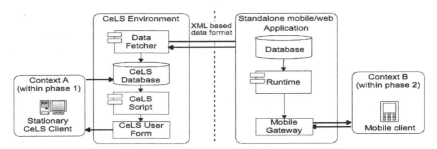

Fig. 1. CeLS integration with standalone applications

2.1   Providing Location Attributes with MoCoLeS

Our initial attempt to integrate an external system with the CeLS platform involved an environment referred as MoCoLeS (Mobile Collaborative Learning System) that supported the implementation of location-based activities enacted by GPS enabled mobile devices [8]. This integration was implemented by a script that combined an outdoor phase with indoor phases performed later on at home in a asynchronous manner. The outdoor phase dealt with identification and documentation of the usability problems in the campus while the indoor phase dealt with tagging the problems reported by the participants and conducting a competition between the contributed artifacts. The data integration process is initialized while CeLS sends to MoCoLeS a request for relevant activity data including unique identifiers. In response, MoCoLeS sends to CeLS data that corresponds to the appropriate elements identifiers (a picture that illustrates the usability problem and GPS data). This migrated data is stored at the CeLS database along with its unique identifiers in order to enable further interaction with it in a later phase. Such activities can include for instance, peer's asynchronous interaction with contributed artifact at the learner's home. This example illustrated the combination of mobile and stationary technologies that can provide support across different physical learning spaces. The kind of data integration illustrated in this example provides CeLS script with new capabilities for supporting collaborative outdoor activities.

2.2   Expanding CeLS for Face to Face Classroom Activities with SMS-HIT

The second example involves the integration of the CeLS platform with SMS-HIT, a PRS based on mobile devices for SMS and web response, designed to enable instructors to prepare and enact interactive class activities [9]. This integration process was implemented and used in a learning activity that combined data collection during a regular classroom lesson via SMS (identifying and declaring one's own negotiation style). Additional phases were later performed at home (responding to various negotiation scenarios according to the self reported style and evaluating peer responses). The CeLS data fetcher requests the application (SMS-HIT) for specific learners' contributions that correspond with the specific data activity according to the unique identifiers manifested in the CeLS script. In response, the SMS-HIT sends back the data contributed by SMS along its unique data identifiers in an XML based data format. The data is migrated to the CeLS database and is allocated according to the unique identifier of the data requestor within the script. This example illustrates an additional effort that enables teachers to design and implement rich and complex pedagogical strategies that include asynchronous, as well as face-to- face class learning activities as interconnected elements. This expansion requires the combined use of various communication technologies, whenever suitable.

3   Identifying Key Elements for Supporting Systems Integration

The integration process between the different technological components used during a cross context learning activity requires a profound analysis of its unique requirements [10]. The implementation of such integration involves planning of several aspects like data and workflow integration. The data level concern the mediation and integration among data that originates from different applications used to support the learning process. Data sets could emerge from a variety of sources, like web and mobile applications or other hybrid types of environments [11].

The data that is exchanged during the integration often involves multiple related blocks of information and may be conveniently organized with XML data structures. The nature of the interrelated contexts along the learning activity requires that the data exchange will contain a detailed description of the activity elements that are manifested along the phase's notation within the script. It should also include their logic, their specific rules and their inter-relation dependencies between the elements along the script. Current learning script standards are challenged with issues of representation of contextual metadata [12].

The integration process refers to the choreography between the involved applications used in different contexts. It also refers to the necessity to provide continuous data flow and activity support across different phases of the pedagogical scenario [13, 14]. Furthermore, it should enable meaningful representation of the created artifacts and their reuse across different phases of the scenario using various technological devices and services.

3.1   Multiple Layer of Interdependencies for Supporting Systems Integration

The integration of distinct systems and applications used along the different learning contexts directly influences the data interoperability that relates to the artifacts produced along the activity phases in relation to previously contributed artifacts from a different context [15]. This also implies the required interdependencies among the different systems layers of the technological environment in which the different artifacts are created within different learning contexts. Fig. 2 illustrates a model that can cope with cross context learning activity.

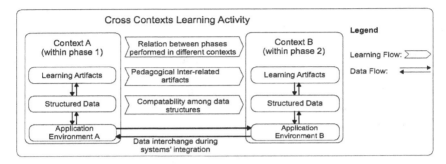

Fig. 2. Cross context systems integration model

The integration process involves multidimensional aspects like social setting, location and compatibility among technological devices involved when dealing with cross context script. This process involves several challenges related to different layers of integration. Interdependency exists between the learning artifacts layers across context that implies on the relation between the data structural layers. The integrated applications within each of the involved contexts participate in the data interchange during the advancement of the activity to the next phase. Furthermore, the involved layers within each context communicate with nearby layers that are part of each block. For example, an artifact is represented in a form of a structured data that is passed from one environment to the other (integrated applications). The receiving application interprets the structured data for the use as a learning artifact. This multi-layer illustration also highlights the challenges involved in cross context learning like the requirements to foster seamless learning by supporting continuous data flow along the activity [15, 16]. Specifically, it also highlights the importance of the interdependencies while dealing with complex data learning structures [6].

The following sections present some of the main elements that emerge from the layers involved in the model. They highlight the importance of some of the stakeholders that are directly involved with the project's goals. We will also deal with the identification of a necessary data format that includes learners' contributions along with unique marked contributions identifiers that support the integration process and finally review the identification process of a collaborative scenario that are supported by systems integration.

3.2   Systems Integration Stakeholders

A typical software development project involves various stakeholders that are related to the project's development process and products. The onion model can provide a depiction of the relevant stakeholders within a series of rings located around the project's core [17]. In our integration projects, the inner rings contain the developers, while the middle ring includes the operators that provide teachers with technical support along the activity. The outer ring includes teachers, learners and also developers (as help desk). From a software engineering perspective, learners and teachers are highly involved with designing and using the pedagogy offered by the integrated cross contexts application. The developers within their different functionalities are involved with the engineering, technical and the maintenance issues related to the integration.

3.3   Identifying and Defining Integration Goals

Goals represent the stakeholders' intentions and could be described at any level of detail [17]. A project may aim at both functional and quality goals. In our integration projects the goals are defined by teachers (as potential users) for enabling efficient classroom face to face or outdoors cross context interactions using technological platforms. The functional goals of the implemented examples aim to support the systems integration across learning contexts. The quality goals reflect the stakeholder's intentions to improve pedagogical practices supported by better architectural and engineering design and development methods for integrated learning technologies.

3.4   Data Format for System Integration

Systems integration is performed by formatted data that is exchanged between the different applications of the technological environments. In our case, the integration contains learners' contributions that were submitted across contexts tagged by unique identifiers. For example, an SMS-HIT application will exchange the SMS message with the CeLS by XML proprietary data format that includes the contributed SMS during the face-to-face interaction marked with unique identifier. The unique identifiers serve for later interaction with learners' artifacts.

3.5   Designing Collaborative Scenarios

Finally, we can identify scenarios that communicate the nature of the situations as they evolve through time in a series of steps [17]. A scenarios design process comprise of stakeholders' involvements, project goals definition and also the development of a data model that enables its functionality. In our case, the scenarios depicted different actions that learners have to perform along the learning paths, as well as a description of interaction between the acting learning environments (different levels of scenarios). Although the specific scenarios of the examples illustrated in this paper are not identical, many commonalities could be identified in terms of the pedagogical activity pattern and the use of CeLS as the main platform that drives the system integration. CeLS enables to describe and enact cross platforms actions by the usage of it APIs and web services. For example, both activities included phases that present learners with request across contexts for contributions, peer assessments and similar type of debriefing.

4   Summary and Future Efforts

We have presented and analyzed the potential and challenges related to systems' integration for supporting pedagogical scenarios performed across different contexts. Specifically, we deal with three different design challenges: conceptual, architectural and engineering. For example, CeLS and SMS-HIT integration requires conceptual design of an activity script followed by software architecture design and implemented by software engineering tools. The challenges include identifying stakeholders, goals, usage scenarios and the data format that is required in order to integrate new elements for the expansion of CeLS´ abilities towards a broader usage context through a successful integration. These challenges will be tackled in our upcoming efforts towards the integration of CeLS with a dedicated web based authoring tool designed to enable teachers to create, enact and reuse outdoor Treasure-Hunt activities performed with mobile devices. This type of activity involves many of the script dimensions that were not addressed in cases described within this paper. The activity may start with a phase performed in CeLS in which learners are challenged to suggest landmarks' descriptions (multimedia and GPS data location) along with riddles that point to these sites of interest. This information will be migrated and used by the Treasure Hunt application in order to define a game that will be enacted via mobile technology. Alternatively, data contributed by the participants during the game via mobile devices may be collected and migrated to the CeLS system and used for later asynchronous activities in class or at home. The integrations described in this paper aim to support multi-phase activities that involve indoor, outdoor and on-line phases addressing different pedagogical approaches such as inquiry learning and collaborative learning. The current integration cases are used as a part of our on-going efforts to develop a CeLS Application Programming Interface (API) for integrating external systems artifacts. We envision that such expansion of the technological infrastructure may foster and support the emergence of new pedagogical ideas.

References

1. Milrad, M., Kohen-Vacs, D., Vogel, B., Ronen, M., Kurti, A.: An Integrated Approach for the Enactment of Collaborative Pedagogical Scripts Using Mobile Technologies. In: Proceedings of the International Conference on Computer Support for Collaborative Learning, CSCL 2011, Hong Kong, China, pp. 681–685 (2011)
2. Dillenbourg, P., Jermann, P.: Technology for Classroom Orchestration. In: Khine, M.S., Saleh, I.M. (eds.) New Science of Learning: Cognition, Computers and Collaboration in Education, pp. 525–552. Springer, Dordrecht (2010)
3. Duncan, D.: Clickers: A New Teaching Aid with Exceptional Promise. Astronomy Education Review 5(1) (2006)
4. Sharples, M., Milrad, M., Arnedillo, S.I., Vavoula, G.: Mobile Learning: small devices, big issues. In: Balacheff, N., Ludvigsen, S., de Jong, T., Lazonder, A., Barnes, S., Montandon, L. (eds.) Technology Enhanced Learning: Principles and Products. Springer, Berlin (2007)
5. Goodyear, P., Retalis, S.: Technology-Enhanced Learning: Design Patterns and Pattern Languages. Sense Publishers (2010)

6. Ronen, M., Kohen-Vacs, D.: Techniques for Fostering Collaboration in Online Learning Communities: Theoretical and Practical Perspectives. In: Pozzi, F., Persico, D. (eds.) Modeling, Enacting Sharing and Reusing Online Collaborative Pedagogy with CeLS, pp. 319–339. IGI Global, Hershey (2011)
7. Pettersson, O., Vogel, B.: Reusability and Interoperability in Mobile Learning: A Study of Current Practices. In: 2012 Workshop on Scalability and Interoperability Dimensions for Mobile Learning Applications (SID-ML 2012); Proceedings of the Seventh IEEE WMUTE 2012, Takamatsu, Japan, March 27-30, pp. 306–310 (2012)
8. Kohen-Vacs, D., Ronen, M., Ben Aharon, O., Milrad, M.: Incorporating Mobile Elements in Collaborative Pedagogical Scripts. In: Proceedings of the International Conference on Computers in Education, ICCE 2011 (2011)
9. Kohen-Vacs, D., Ronen, M., Bar-Ness, O.: Integrating SMS Components into CSCL Scripts. In: 7th IEEE International Conference on Wireless, Mobile & Ubiquitous Technologies in Education (WMUTE 2012), Takamatsu, Japan, March 27-30 (2012)
10. Perez-Sanagustin, M., Emin, V., Hernandez-Leo, D.: Towards the design of learning scenarios combining activities across multiple spaces. In: Across Spaces Workshop, the 6th European Conference on Technology Enhanced Learning, Palermo, Italy, pp. 9–12 (2011)
11. Halevy, A.: Why your data won't mix. Queue 3(8), 50–58 (2005)
12. Stegmann, K., Wecker, C., Harrer, A., Ronen, M., Kohen-Vacs, D., Dimitriadis, Y., Hernandez-Leo, D., Fernandez, E., Asensio-Perez, J., Fischer, F.: How can current approaches to the transfer of technology-based collaboration script for research and practice be integrated? In: Spada, H., Stahl, G., Miyake, N., Law, N. (eds.) Connecting Computer-Supported Collaborative Learning to Policy and Practice: CSCL2011 Conference Proceedings volume III — Community Events Proceedings. International Society of the Learning Sciences, pp. 1103–1110 (2011)
13. Lorenzo, G.D., Hacid, H., Young Paik, H., Benatallah, B.: Data integration in mashups. SIGMOD Record 38(1), 59–66 (2009)
14. Oser, F.K., Baeriswyl, F.J.: Choreographies of teaching: Bridging instruction to learning. In: Richardson, V. (ed.) Handbook of Research on Teaching, 4th edn., pp. 1031–1065. American Educational Research Association, Washington (2001)
15. Kohen-Vacs, D., Ronen, M., Bar-Ness, O., Milrad, M., Kurti, A.: Integrating Collaborative and Mobile Technologies for Fostering Learning about Negotiation Styles. Submitted to the 19th Internation. Conference on Computers in Education (2012)
16. Chan, T.-W., Roschelle, J., His, S., Kinshuk, S.M., Brown, T., Patton, C., Cherniavsky, J., Pea, R., Norris, C., Soloway, E., Balacheff, N., Scardamalia, M., Dillenbourg, P., Looi, C.K., Milrad, M., Hoppe, U.: One-to-one technology-enhanced learning: An opportunity for global research collaboration. Research and Practice in Technology Enhanced Learning 1(1), 3–29 (2006)
17. Alexander, I., Beus-Dukic, L.: Discovering requirements: how to specify products and services. Wiley, Chichester (2009)

Tangible and Wearable User Interfaces for Supporting Collaboration among Emergency Workers

Daniel Cernea[1,3], Simone Mora[2], Alfredo Perez[2], Achim Ebert[1], Andreas Kerren[3], Monica Divitini[2], Didac Gil de La Iglesia[3], and Nuno Otero[3,4]

[1] University of Kaiserslautern, Germany
{cernea,ebert}@cs.uni-kl.de
[2] Norwegian University of Science and Technology, Norway
{simone.mora,perezfer,monica.divitini}@idi.ntnu.no
[3] Linnaeus University, Sweden
{andreas.kerren,didac.gil-de-la-iglesia,nuno.otero}@lnu.se
[4] University of Minho, Portugal

Abstract. Ensuring a constant flow of information is essential for offering quick help in different types of disasters. In the following, we report on a work-in-progress distributed, collaborative and tangible system for supporting crisis management. On one hand, field operators need devices that collect information—personal notes and sensor data—without interrupting their work. On the other hand, a disaster management system must operate in different scenarios and be available to people with different preferences, backgrounds and roles. Our work addresses these issues by introducing a multi-level collaborative system that manages real-time data flow and analysis for various rescue operators.

Keywords: Wearable tangible device, collaborative crisis management.

1 Introduction

Humans, despite technological and scientific advances, are still vulnerable in the face of natural disasters. It is therefore essential to provide effective management and quick aid in such scenarios [8,15]. Providing up-to-date data, ensuring a constant flow of information, organizing and coordinating rescue units and reaching the people in need are the core factors for ensuring disaster management and offering quick help. This paper presents an exploratory design study on tangible user interfaces for improving coordination in crisis management.

Designing novel Disaster Management Information Systems (DMIS) poses unique challenges [1,2]. Multiple publications have focused on interaction techniques for crisis management systems, capturing vital aspects in the areas of multitouch [3, 17] or gesture interaction [1,4], with a special emphasis on map-based approaches. At the same time, solutions have been devised that aid the cooperation and interaction of disaster managers and unit operators in the settings of a mobile command post connected to mobile devices [18]. Still, while mobile devices like tablets and smart phones would seem ideal, the need for additional information about the environment

V. Herskovic et al. (Eds.): CRIWG 2012, LNCS 7493, pp. 192–199, 2012.

[6,7] and specific operation conditions has lead to scenario-fitted approaches, where field operators employ handheld [5] and wearable devices [14].

We aim at exploring how tangible interaction impact on crisis management and we propose a prototypical system implemented by a tabletop interface for team coordinators and disaster managers sitting in a control room, and by a wearable interface attached to each field agent's forearm. In the following sections, we describe the features and functionality of our tangible collaborative system. Next, we focus on the evaluation of our system by a group of experienced rescue workers and discuss our findings. Finally, we conclude highlighting the major findings, their implications and plans for future work.

2   User Studies and Scenario

In this section we highlight a scenario for the proposed crisis management system. The scenario has been developed building on observations and interviews with emergency workers performed during a three-day simulation of a massive disaster held in Italy in 2011. Scenarios included flooding, earthquake and a massive car jam. Rescue workers were deployed to find and rescue persons (i.e. actors impersonating injured persons) in a physical environment that resembled a real disaster; team coordinators and disaster managers were directing operations from a control room. Teams included rescue units, civil protection, police, responder for hazardous and chemical contamination, dog rescue units. One of the paper's authors shadowed workers assigned to different roles during the three days in order to gain an understanding of procedures and technology in use for coordination during a crisis response.

Results from the study show that agents still largely rely on handheld transceivers (i.e. walkie-talkies) to communicate among each other and with the team coordinators. Once the rescue and management operations are underway, the field agents are given instruction by coordinator through radio broadcasts. At the same time, field workers have to communicate back information like their position, environmental data (temperature, humidity, air quality) in a half-duplex communication. As this can be only done in a qualitative way, often their information can get biased or distorted [13]. Additionally, the units on the field need to remember and execute the tasks and commands assigned to them by the coordinators. Meanwhile, coordinators in meeting rooms need to transcribe the radio communication, as well as annotate and update on a map the positions of the teams and data they have collected. Building on results from the study we have developed a scenario to show how the use of tangible and wearable technology might impact on the work practice.

The Scenario
Scene: EM Coordinators in a Mobile command Center - Disaster managers activate an emergency response gathering around a tabletop in a mobile unit (Figure 1, left). They explore a map of the disaster and decide where to deploy the emergency units. There are different units depending on the specific disaster to consider (e.g. flood, fire, earthquake, etc.).

Scene: EM Workers on the Disaster Scene - Workers arrive to the crisis scene. Wearable devices are consulted in order to identify context information (the place they have been assigned, noise level and temperature) and tasks to carry out (Figure 1, right). A worker has received a notification requesting his reallocation in a different sector. Once having performed the task he acknowledges the conclusion by interacting with his wearable device.

Fig. 1. The dimensions of collaboration supported by the emergency management system

3   Design and User Interaction

The proposed solution we have designed consists of two main elements with wireless capabilities for information sharing. These elements are a *tabletop unit* used to manage and coordinate the different field units, and a *wearable unit* in the form of a wristlet to be used by workers on the field.

3.1   Tabletop Unit

The proposed Mobile Command Center (MCC) has the role of supporting the decision making process of the unit coordinators. The tabletop allows multiple users to interact with the map of the affected region, as well as gather and analyze data a constant data stream (real-time information from wearable devices). Furthermore, the tabletop can receive and display new information about the crisis situation as it pours in from the public to the rescue services (e.g. dispatcher). At the same time, coordinators have the possibility to independently send messages to the field units to inform them about new developments or give instructions on how to proceed.

The initial step for managing a crisis is the registration of the event in the MCC. To do so, colored marker objects (Figure 2) are used, which are meant as an efficient and intuitive way to manipulate the location and type of the reported disaster. The color of the marker encodes the type of the disaster, allowing other rescue services to be automatically informed.

Furthermore, the coordinators can interact with the marker objects to update the information about the disaster. Once the changes to the event are made, the marker object can be removed from the tabletop to avoid occlusion. Placing the marker on the same event site allows users to customize or delete the corresponding disaster information. After an event is registered, the MCC uses its wireless Internet connection to query the database of the rescue services, in order to get additional information about the rescue efforts. Aiming to support readability and collaboration, the users have the possibility to rotate any text or marker.

Fig. 2. Tabletop running the MCC system. Colored marker objects enable the user interaction with information from the field.

One of the most important tasks of the MCC is the communication and coordination of the units on the field. In this sense, each wearable tailored device sends a constant stream of data to the tabletop via a wireless Internet connection. This information is evaluated at the MCC and displayed in real-time for each unit. The collected environment data can be collaboratively and interactively visualized. Coordinators can also access a priority list for sent messages that highlights all the tasks and their current status (received, confirmed or executed).

3.2   Wearable Unit

The wearable device is to be worn on the field worker's arm (Figure 3). User interaction is supported by a LCD color display and a proximity-activated button located on the armband that holds the device. Interaction with the device is designed to disrupt rescue operation as little as possible: high-contrast colors have been chosen in order to enhance screen readability under direct sunlight, while the proximity button can be activated even wearing gloves.

Once activated the device start displaying the following information: GPS coordinates and ground speed, environmental temperature, noise level, the task that the user is assigned to (pre-defined on the tabletop unit), a green/red bar indicating whether the assigned task has been completed or not.

GPS and environmental data are also transmitted to the tabletop unit via a wireless connection. We designed the device to be based on modules so different type of sensor and network adapters can be adopted to address the precise disaster need. A proximity-activated button is located on the device armband. By brushing the armband the user can acknowledge the coordination unit that a task has been completed. The status bar on the display turns green to confirm to the user that the task-completion message has been sent to the tabletop unit and the device is ready to receive a new task.

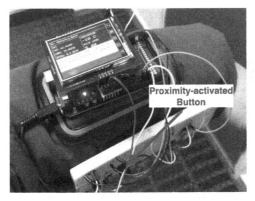

Fig. 3. Proximity-activated button is positioned on the armband. The device notifies the user about a new task received with audio and haptic feedbacks.

4   Architecture and Implementation

4.1   Tabletop Unit

The MCC system that corresponds to the different coordinators of the disaster management efforts, was implemented on a MultiTouch Cell tabletop[1]. The core ideas behind this system include the support for collaboration between multiple unit coordinators, as well as the mobility of a vehicle-mounted tabletop as a communication and management platform. The software running on the tabletop is written in Adobe Flash ActionScript 3. Additionally, the Google Maps API for Flash[2] has been used to allow the manipulation of interactive maps.

4.2   Wearable Device

The first prototype of the wearable unit has been developed using an Arduino Mega board[3] as central module. Both hardware and software have been developed for this work. User interaction is assisted by a 2.5" LCD display, sound and haptic notifications are provided by a buzzer and a small vibration motor. The user can send input to

---

[1] Multitouch Cell. http://multitouch.fi/products/cell/
[2] Google Maps API for Flash.
  https://developers.google.com/maps/documentation/flash/
[3] Arduino Mega. http://arduino.cc/en/Main/ArduinoBoardMega

the device using a proximity-activated button working in infrared light, which allows for use also wearing gloves. Location tracking is enabled by a 66 channels GPS chip, which senses the user's location 5 times per seconds. Network communication is available via an XBee S1 adapter, according with specifications[4] it provides a 120mt communication range with low power demand, the Xbee socket on the board is also compatible with other adapters, as for example the Xbee-PRO 868 capable of communication range up to 80km (outdoor, line-of-sight). Tests have shown that the system can be powered up to 90minutes depending on environmental temperature and the strength of the GPS signal. Increased autonomy can be provided with arrays of 9V batteries, trading autonomy against device weight and size. The software running on the MEGA board has been developed using the Arduino 1.0 SDK, the TinyGPS library[5] has been used to parse data from the GPS chip.

5   Formative Evaluation

To gather feedbacks, we recorded a video to illustrate the scenario and usage of the prototypes. After having watched the video, seven testers were asked to fill in a questionnaire using a Likert-type scale and open questions. Questions were elaborated around five areas considered important for assessing the technology acceptance of the system: scenario and problem analysis, system architecture, user interaction (overall), wearable device interaction, tabletop device interaction. We showed the video during two videoconference sessions and collected results from the questionnaire as well as informal discussions. Participants of our evaluation were both IT technical staff working for emergency response forces and the EM workers themselves. Results and implication for design are reported below.

Scenario and Problem Analysis - Results from the evaluation show a good user acceptance of the system, the workers clearly stated their interest in using the system frequently. The scenario we are addressing is also acknowledged as well grounded since it often occurs during emergencies.

System Architecture - The different devices and functionalities of the system were properly interpreted by the interviewees, and the overall functionalities were considered coherent. The data types the system is sensing and displaying (location, noise, temperature) to enhance collaboration are considered meaningful during crisis management. Moreover, the users have shown a strong interest in allowing for tailoring the system by adding more sensing capabilities to meet certain crisis scenarios.

User Interaction - Although the efforts made in creating a user-friendly design, the system is still considered somehow complex and we need to do more work on both the wearable and the tabletop prototype. On one side the system is considered easy to use, but a need for a tutor or course to get started with it is suggested. None of the interviewees considered it feasible for the user to set-up the system and to configure networking functionalities.

---

[4] XBee. http://www.digi.com/xbee/

[5] TinyGPS, http://arduiniana.org/libraries/tinygps/

Wearable Device Interaction – The wearable unit is currently considered too big and heavy to be successfully employed in real rescue operations; an improved hardware design and the production of custom electronic parts could drastically reduce the device dimensions and weight. Also further user studies and design workshops should be carried out in order to choose where to position the wearable unit on the user's body (*"Forearms and hands are needed to be free for movements and to raise weights"*). Feedbacks reveal that on the field, users might want to have more control on data collected by sensors, for example by being able to set the granularity or the sample frequency. Some comments suggested to allow the device to display sensor data from nearby units worn by colleagues: *"It would be useful to have the data displayed for all units, all the time"*; this is intended to give to the field agents extended awareness of the local environment and potential dangers.

Tabletop Device Interaction - The tabletop unit is considered helpful, but interviewees stated the need to improve both data visualization and interaction modalities. Discussions with the interviewees have shown that users would benefit from an extended range of physical artifacts and gestures to interact with the user interface. Also we should investigate additional visualization metaphors (heat maps, tag clouds) in presence of a huge amount of data and offer further solutions for avoiding occlusions.

6 Conclusions

In this paper we presented a scenario-based development of a distributed tangible system to support disaster management. The potential of the system is highlighted in a formative evaluation that involved emergency workers and IT consultants with expertise in IT systems for crisis management. The preliminary evaluation acknowledged the system as useful, although future works are needed to improve the design particularly in the usability area. In the future, we plan to build on the evaluation results and to involve emergency workers in participatory design sessions in order to elicit additional information and inform the development of new prototypes.

References

1. Wang, F., Wen, R., Zhong, S.: Key Issues in Mapping Technologies for Disaster Management. In: Proc. 2nd International Conference on Information Engineering and Computer Science (ICIECS), pp. 1–4 (2010)
2. Frassl, M., Lichtenstern, M., Khider, M., Angermann, M.: Developing a System for Information Management in Disaster Relief - Methodology and Requirements. In: Proc. 7th International ISCRAM Conference (2010)
3. Nebe, K., Klompmaker, F., Jung, H., Fischer, H.: Exploiting new interaction techniques for disaster control management using multitouch-, tangible- and pen-based-interaction. In: Jacko, J.A. (ed.) HCII 2011, Part II. LNCS, vol. 6762, pp. 100–109. Springer, Heidelberg (2011)

4. Artinger, E., Coskun, T., Schanzenbach, M., Echtler, F., Nester, S., Klinker, G.: Exploring Multi-touch Gestures for Map Interaction in Mass Casualty Incidents. In: 3. Workshop zur IT-Unterstützung von RettungskräftenimRahmen der GI-JahrestagungInformatik (2011)
5. Mecella, M., Angelaccio, M., Krek, A., Catarci, T., Buttarazzi, B., Dustdar, S.: Workpad: an Adaptive Peer-to-Peer Software Infrastructure for Supporting Collaborative Work of Human Operators in Emergency/Disaster Scenarios. In: Proc. International Symposium on Collaborative Technologies and Systems (CTS 2006), pp. 173–180. IEEE Computer Society (2006)
6. Fischer, C., Gellersen, H.: Location and Navigation Support for Emergency Responders: A Survey. IEEE Pervasive Computing 9(1), 38–47 (2010)
7. Lorincz, K., Malan, D.J., Fulford-Jones, T.R.F., Nawoj, A., Clavel, A., Shnayder, V., Mainland, G., Welsh, M., Moulton, S.: Sensor Networks for Emergency Re-sponse: Challenges and Opportunities. IEEE Pervasive Computing 3(4), 16–23 (2004)
8. Bergs, J., Naudts, D., Van den Wijngaert, N., Blondia, C., Moerman, I., Demeester, P., Paquay, J., De Reymaeker, F., Baekelmans, J.: The ADAMO project: Architec-ture to support communication for emergency services. In: 8th IEEE International Conference on Pervasive Computing and Communications Workshops (PERCOM Workshops), pp. 382–387 (2010)
9. Tierney, K., Sutton, J.: Cost and culture: Barriers to the adoption of technology in emergency management. Rescue Research Highlights (2005)
10. Manoj, B.S., Baker, A.H.: Communication challenges in emergency response. Commun. ACM 50, 51–53 (2007)
11. Scott, S.D., Grant, K.D., Mandryk, R.L.: System guidelines for co-located, col-laborative work on a tabletop display. In: Proc. of ECSCW 2003, pp. 159–178. Kluwer Academic Publishers (2003)
12. Tang, A., Tory, M., Po, B., Neumann, P., Carpendale, S.: Collaborative coupling over tabletop displays. In: Proc. SIGCHI Conference on Human Factors in Computing Systems (CHI 2006), pp. 1181–1190. ACM (2006)
13. Monares, A., Ochoa, S.F., Pino, J.A., Herskovic, V., Rodriguez-Covili, J., Neyem, A.: Mobile computing in urban emergency situations: Improving the support to firefighters in the field. Expert Syst. Appl. 38(2), 1255–1267 (2011)
14. Curone, D., Dudnik, G., Loriga, G., Luprano, J., Magenes, G., Paradiso, R., Tognetti, A., Bonfiglio, A.: Smart Garments for Safety Improvement of Emergency/Disaster Operators. In: Proc. the 29th Annual International Conference of the IEEE EMBS Cité Internationale, Lyon, France, August 23-26 (2007)
15. Meissner, A., Zhou, W., Putz, W., Grimmer, J.: MIKoBOS - a mobile information and communication system for emergency response. In: Proc. 3rd Intl.Conference on Information Systems for Crisis Response and Management, pp. 92–101 (2006)
16. Lachner, J., Hellwagner, H.: Information and Communication Systems for Mobile Emergency Response. In: UNISCON 2008, pp. 213–224 (2008)
17. Bader, T., Meissner, A., Tscherney, R.: Digital map table with Fovea-Tablett: Smart furniture for emergency operation centers. In: Proc. 5th International Conference on Information Systems for Crisis Response and Management, pp. 679–688 (2008)
18. Piazza, T., Heller, H., Fjeld, M.: CERMIT: Co-located and Remote Collaborative System for Emergency Response Management. In: Proc. SIGRAD 2009, Visualization and De-sign, Goeteborg, Sweden, p. 12 (2009)

Contextual Analysis of the Victims' Social Network for People Recommendation on the Emergency Scenario

Sírius Thadeu Ferreira da Silva, Jonice Oliveira, and Marcos R.S. Borges

PPGI/IM – Graduate Program in Informatics, Institute of Mathematics (IM)
Federal University of Rio de Janeiro (UFRJ)
Rio de Janeiro – RJ – Brazil
sirius@ufrj.br, {jonice,mborges}@dcc.ufrj.br

Abstract. The growing use of mobile devices by the population and the high popularity of the social media in current society, such as Facebook and Twitter, produces more and more information, plenty of them with contextual data. One of the major obstacles to the emergency response team during the response phase of emergency management is to obtain information that could lead to solving a particular situation involving emergency victims. In this paper we present a proposal which aims to collect information from social media and mobile devices, identify the contextual information and analyze them to indicate people who could help in the identification of victims. This work focuses on identifying the social network of victims and look for people who can provide important and reliable information about them, thus assisting the emergency team in its work. We use this contextual information to improve the recommendation process, identifying people with high degree of closeness.

Keywords: Social Networks, Emergency Response, Recommendation.

1 Introduction

The widespread use of mobile devices and the boom of the social media that occurs in our society suggests the use of these resources to address problems and unexpected situations that evolve over time and put lives in danger. This work aims to analyze the information found in mobile devices and social media to help in emergency situations. We are addressing the lack of reliable information about emergency victims issue.

The work main idea is to infer the closest people in the victim's social network and recommend them, thus helping the emergency team to collect important and trusty information about the victims. We think that identifying the victim closer friends who are nearby the disaster local can improve the recommendation process.

This paper is organized as follows: section two presents a general view on the emergency management issue and the problem addressed. The third section describes our solution proposal, showing the contextual model, the envisioned architecture, the criteria used during the implementation and defining our scope, besides works we were inspired by. The fourth section presents conclusions and suggests future work.

V. Herskovic et al. (Eds.): CRIWG 2012, LNCS 7493, pp. 200–207, 2012.
© Springer-Verlag Berlin Heidelberg 2012

2   The Emergency Management Issue

The emergency area is nebulous and there are a large amount of topics and definitions that can be found in literature. But in general, emergency can be defined as a sudden event that calls for immediate measures to minimize its consequences [1].

Emergency Management is the process by which the uncertainties that exist in potentially hazardous situations can be minimized and public safety maximized. The goal is to limit the costs of emergencies or disasters through the implementation of a series of strategies and tactics reflecting the full life cycle of disaster, i.e., preparedness, response, recovery, and mitigation [2]. The work presented here has its focus in the response phase of the emergency management's life cycle.

The response phase is the most complex and therefore the most studied of them all. Factors that deal with complexity are the unpredictability and the speed of events, the number of involved people, shortage of time to make decision and act as planned, the unavailability of resources and some uncertainty about the situational awareness.

One of the biggest challenges of the emergency response team is to obtain reliable information on accident victims who are missing or unconscious. Imagine an emergency scenario where a landslide occurred on a hillside community. As a result of this accident, there was the burial of several houses in the area. Some of the people who lived in this community were found, but others are missing or unconscious. How could the emergency response team act to find information about these missing or unconscious people? Most likely they would have to personally ask the people involved in this emergency scenario for more information about these victims.

3   Proposal: Contextual Analysis

Based on the above issue, the purpose from this paper is to present an effective and efficient process that starting with the missing or unconscious victim's name can build his/her social network through collecting the contextual information contained in the mobile devices of people nearby the emergency scenario and the social media.

This solution aims to provide means of minimizing the work of the emergency response team in collecting information about victims of an accident, while trying to maximize the reliability of this information (recommending only those people that are most relevant to this process of information collection, i.e., the closest people to the victim), thus speeding up the victim's rescue procedure. Our main objective is people recommendation based on the social relevance among them and the victim.

Fig. 1. Data treatment and social relevance processing

Table 1. Variables related to the social distance

Variable	Meaning
Relationship Distance (RD)	Number of friendship links between two persons. The distance is 1 for direct friends, 2 for the friends of friends, 3 for friends of friends of friends, and so on.
Relationship Type (RT)	Kind of tie or link among people. We divided it in: LLR (Long Live Relationship, such as parents or spouse), Friend, Family (be member of), Co-worker and Acquaintance.
Relationship Weight (RW)	We believe that some types are stronger than others. So, we define some weights to these types: LLR – 5; Friend – 4; Family (be member of) – 3; Co-worker – 2; Acquaintance – 1.
Number of Connections (NC)	Frequency of communication among people. We get it by number of SMS and calls (in mobile phones) and interaction in social media (as comments, likes, content sharing, messages, and others).

3.1   Social Relevance

The main activity of this proposal is the identification of the social relevance of users. For this, we follow the process described in Fig.1. As data sources, the interactions on Facebook, Twitter, phone calls, e-mails and SMS texting are used. We calculate the social relevance using some parameters which are presented in Table 1.

Our Social Relevance (SR) is $SR = RW * NC$.

If RD = 1, we use RW as in Table 1. If not RW = 1/RD. The Social Distance (SD) is inversely proportional to the Social Relevance (SR). The more the social relevance (of a person, related to other) is higher, closer they are (smaller social distance). The calculus of the social relevance tries to identify the affinity of the victim with someone else. In the recommendation process, the next step is ranking and identifying people with high social relevance (or small social distance).

The Relationship Distance term was based on the small-world experiment [3], which examined the average path length for social networks of people in the United States. This research suggested that human society is a small world type network characterized by short path lengths. This work showed that the world is increasingly interconnected by stating that only five intermediaries (on average) would be enough to connect any two randomly chosen individuals, regardless of where they lived.

The Relationship Type term represents the interpersonal ties, which are defined by mathematical sociology as information-carrying connections between people. Kapferer postulated the existence of multiplex ties, characterized by multiple contexts in a relationship [4]. Multiplexity is the overlap of roles, exchanges, or affiliations in a social relationship [5]. Therefore, the Relationship Weight term in these multiplexity cases is adjusted to use the greatest weight of the multiplex ties.

The Number of Connections term is based on the "strength" of an interpersonal tie, i.e., a linear combination of the amount of time, the emotional intensity, the intimacy (or mutual confiding), and the reciprocal services which characterize each tie [6]. We calculate it by assigning different weights for each kind of interaction between two persons: likes and citations (1), comments and content sharing (2), messages (3), SMS (4) and calls (5). We make a weighted average using these data, plus the calls total duration for giving a boost, since the conversation time can be an intimacy indicator.

3.2   The Overall Process

Our proposal has a three tier client-server architecture. The first tier is the web server, a repository for information exchange and research users' data. The second tier is the command and control server, a mobile server represented by the system installed on all mobile devices used by the emergency response team. This node sends queries looking for people that can help in an emergency, like citizens who can provide information about a victim. The third and last tier is the mobile client, represented by the system installed in all mobile devices used by the population. This node provides important information, such as if the user is part of the social network of the victim.

When the user (ordinary citizen) installs our application on his/her mobile device (Fig. 2a), it registers itself on the web server, providing some information (Fig. 2b). The web server stores the user information and search more useful information about him/her on the social media (Facebook) (Fig. 2c), forming an users' information cache (Fig. 2d). All his/her contacts in mobile phone is sent to be processed, as also frequency and duration of calls and SMS. These data is processed to identify the social distance among users and frequently updated, as described in section 3.1.

During an emergency, the responsible team (Command And Control Server) will seek information about the victims (Fig. 2e). The command and control server sends requests for mobile clients within a pre-defined action radius (Fig. 2f). The mobile clients search the user's social network after the query matching (victim name) (Fig. 2g). In a match, the client responds to the command and control server indicating that the user knows the victim and the degree of the relationship importance. If none of the mobile clients in the vicinity of the emergency site responds the query, the web server is reached for the contextual analysis of the victim's social network (Fig. 2h).

The web server then checks the information about the victim, first seeking the data in the previously constructed cache (Fig. 2i). In this cache we have the social distance from a person to his/her friends or acquaintances. The web server returns a response to the command and control server with the identified victim's social network. Closest users to the victim are recommended to assist the emergency responders, following the pre-established selection criteria. The architecture of this proposal is in Fig. 2.

3.3   Privacy Issues

Wagner et al. [7] shows that in social networks, the participants are more diligent and careful about sharing other people's information compared to when sharing their own. Novice users can be privacy insensitive, not comprehending how the information is revealed. But then they recognize the importance of controlling the availability of the data through mechanisms such as disabling the service. Usefulness of information sharing services was acknowledged in more stressful situations as in crisis scenarios in general. In such situations, information usefulness outweighs privacy concerns.

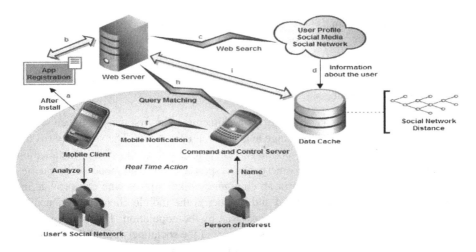

Fig. 2. Proposed solution architecture

We are aware of the research implications regarding the users privacy. We know that because our solution needs to collect information about the user contacts, calls, messages exchanged, among others, it is important that we address this privacy and security issue. Initially, the contract to use the solution will contain a disclaimer saying that we will use the data collected only in emergencies situations, and that only the emergency response team will have access to these data. In addition, the system will include options to I) provide the data automatically (always provide the data), II) provide the data only in certain time periods (configured by the user) and III) provide the data manually (requesting permission to the user before providing the data).

One of the advantages of the proposed solution is the fact that it collects the data and make them available to emergency responders proactively. However, we will provide a reactive option, which reduces the speed of the whole process. Only a list of the contacts names that matches the query will be provided to the emergency response team. We will not provide the contacts phone numbers. All data provided by the application will be encrypted before being sent to the emergency response team.

3.4   Implementation

We implemented and tested the data collection from the user's mobile device. The prototype takes the victim name and searches the user's contacts. For each matched entry, we check the amount of calls, the last call date, the calls (incoming/outgoing) total duration and the exchanged messages (SMS/MMS). These data will be used to calculate the social closeness index of the contact (in relation to the device owner).

We also use data from social media in this calculation. After creating the contacts' social closeness index, the user's mobile device sends to the response team a list of contacts that matched the query, ordered according to the social closeness ranking (from the most relevant to the least relevant) along with some contact extra data (the more important one being the contact photo, if present in the mobile device).

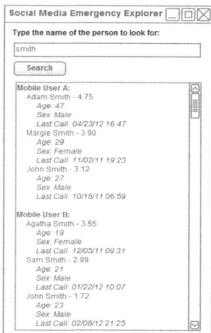

Fig. 3. Prototype shows result list for the query looking for the name "Paulo"

Fig. 4. Result list of nearby mobile devices for the query looking for the name "Smith"

All of these design decisions were taken in order to provide means of avoiding the homonyms problem. Imagine searching for a victim name in our solution and it indicates multiple users with high levels of social closeness to the supposed victim. But which of these users actually knows the victim, and not a homonym? Because of these issues we found that it's necessary to send these extra data, so that emergency responders can quickly eliminate these "false positives". Without these artifices, our solution could easily disrupt the work of the response team rather than facilitating it, making the responders look after people who actually do not know the real victim.

Through the contacts ancillary data analysis, the response team can confirm that he/she is really the victim we are looking for. For example, comparing the victim apparent age and the contact age, making sure the victim and the contact sex are the same, analyzing the last call date/time – if the user has communicated with the contact hours/minutes after the accident, then we can discard this contact as being the victim.

3.5   Preliminary Evaluation

We did some unit tests with the mobile devices data collection functionality and initial results indicate that the most contacted people could easily provide details about the victim in question. Fig. 3 shows one of these unit tests results. We can identify that of the many contacts that match the victim name, one in particular has much relevance to the mobile device owner. So, if this contact is confirmed as the victim himself, the mobile device owner could be convened to provide information about him.

The prototype does not send data to another mobile device, but Fig. 4 shows how the screen would display the list of users and their contacts who matched the query. Our experiment will gather a group of people within a social network and elect one of them as the victim. Then, we run the prototype on each mobile device searching for the victim name. From each result list (which should include victim homonyms), we check which users are most indicated to help providing information about the victim. The person elected as victim will have to indicate whether the trial actually meets the reality (i.e., the prototype ranking named the right person as being the closest to him).

3.6   Related Work

In all studies below, the focus is get any information about the victim, such as his/her location, health status or he/she is alive. They all behave the same: offering an unified repository of information about victims of a particular emergency. Our work focus is built in a different basis. We do not aim to provide the information itself, but we focus on the recommendation of persons to provide reliable information about the victim.

The Katrina PeopleFinder Project [8] was established in response to dozens of groups collecting lists of "lost and safe" for the people affected by Hurricane Katrina. As happened in the aftermath of September 11, a plethora of sites had assembled lists of survivors and missing persons. This project created a system to enter data according to a standard format, and aimed at other sites collecting this kind of information, encouraging them to use the same database, avoiding effort duplication.

The Google Person Finder [9] is an open source web application that provides message boards and records for survivors, family and loved ones affected by natural disaster and seek to provide information about the status and location of people. The system database and API are based on the People Finder Interchange Format. It is embedded in a Crisis Response page, which contains other disaster's tools, such as satellite photos, local shelter, road conditions and information about power outage.

4   Conclusions and Future Work

In this paper we presented an ongoing work that aims to create a better person recommendation process which can be applied in the emergency management field. This process can recommend the closest people to an emergency victim so that they can help the emergency responders with valuable trusty information about the victim. It analyzes the emergency victim's social network and other contextual information inside social media or mobile devices in search of the closest people to the victim.

We are improving the prototype by implementing the social media data collection funcionality and building the social closeness heuristcs. As for the social closeness algorithm and the ranking contruction, we already have done something, but still have to test it. We are revising the proposed architecture in order to improve the data cache, periodically updating it with data collected from the mobile device and social media.

As we plan maintaining a long user data history (i.e., since their inclusion in social media sites or the first mobile device call ever) to create a social closeness index as accurate as possible, we already identified the internet data exchange as being our solution bottleneck. To avoid or minimize this problem, we are thinking of changing the architecture to a two tier (discarding the web server) or pre-calculating the social closeness index at the web server, minimizing the amount of data exchanged through the internet. Each approach has its pros and cons that deserve serious attention.

We recognize that given the long window of time that it is used to estimate the social closeness index, any interaction made years ago has the same weight as a recent one. This can cause trouble in situations where exists individuals with equal social closeness index. So far, for a tie-braker, we use the last call date ancillary data. Although, in future implementations, we can elaborate a social closeness heuristcs that gives different weights for interactions based on its time periods. The chosen weight values for the interaction types where arbitrary, but we plan to evaluate them so we can estimate better values and, in the future, allow weight parameterization.

As a work in progress, we may still take additional actions in order to improve the treatment of the privacy and security issues for the user's information collected. In relation to experimentations, our prototype preliminary evaluation indicates that our social closeness heuristcs leads to indicating the closest people of the user social network, with a given name. Still, we have to run more deep experiments with distinct groups of people, some syntetic data tests and field tests in cooperation with brazilian emergency organisations to see if our proposed solution is really market viable. These experiments should simulate our solution usage during an emergency, so we can evaluate our recomendation process and the prototype usability in general.

References

1. United Nations, Department of Humanitarian Affairs: Internationally Agreed Glossary of Basic Terms Related to Disaster Management. United Nations, Geneva (1992)
2. Drabek, T.E.: The Social Dimensions of Disaster. Federal Emergency Management Agency, Emergency Management Institute, Emmitsburg, MD (1996)
3. Milgram, S.: The Small World Problem. Psychology Today 1(1), 61–67 (1967)
4. Kapferer, B.: Norms and the Manipulation of Relationships in a Work Context. In: Mitchell, J.C. (ed.) Social Networks in Urban Situations. Manchester University Press, Manchester (1969)
5. Verbrugge, L.M.: Multiplexity in Adult Friendships. Social Forces 57(4), 1286–1309 (1979)
6. Granovetter, M.S.: The Strength of Weak Ties. The American Journal of Sociology 78(6), 1360–1380 (1973)
7. Wagner, D., Lopez, M., Doria, A., Pavlyshak, I., Kostakos, V., Oakley, I., Spiliotopoulos, T.: Hide and seek. In: Proceedings of the 12th International Conference on Human Computer Interaction with Mobile Devices and Services, Lisbon, p. 55 (2010)
8. Murphy, T., Jennex, M.E.: Knowledge Management Systems for Hurricane Katrina Response. In: Proceedings of the 3rd International ISCRAM Conference, Newark, p. 615 (2006)
9. Google, Inc. Google Person Finder, http://google.org/personfinder

Matchballs – A Multi-Agent-System
for Ontology-Based Collaborative Learning Games

Sabrina Ziebarth, Nils Malzahn, and H. Ulrich Hoppe

Universität Duisburg-Essen
{ziebarth,malzahn,hoppe}@collide.info

Abstract. Computer games are currently one of the computer science applications with the highest amount of users. The "serious gaming" approach tries to use the attraction (i.e. the fun factor) of such media not only for entertainment purposes, but also to convey serious content at the same time. Serious games have been established in vocational and advanced training over the last years and have a big potential for informal further vocational training. This paper presents a multi-agent-architecture for collaborative, serious and casual games. The focus is on casual games, since these are known to be small games with a high potential for frequent gaming by people of various social and educational background. To be flexible concerning the learning domain an ontology-based approach has been used. The ontology may easily be exchanged to adapt the game to another domain. Furthermore, the data created in the games can be used in a "wisdom of the crowd" approach to enhance the ontology. To test our architecture, an ontology on food safety and hazardous material regulations was created and the game was evaluated with a group of technician students of a professional training academy.

Keywords: CSCL, Multi-Agent-Architecture, Serious Games, Games with a Purpose, Ontologies.

1 Introduction

This paper presents a multi-agent-system for games, which are collaborative, serious and casual as well as have a purpose. The single aspects of this approach are explained in detail in the following sections.

1.1 Serious Game

Computer games are currently one of the computer science applications with the highest amount of users. The "serious gaming" approach tries to use the attraction (i.e. the fun factor) of such media not only for entertainment purposes, but also to convey serious content at the same time. Serious games are often used to virtually train situational behavior like conflict resolution or firemen training or to implicitly transport some knowledge that would not be transferred easily otherwise, because it is too abstract (e.g. nutritional education for young diabetes patients) [1]. They have been

V. Herskovic et al. (Eds.): CRIWG 2012, LNCS 7493, pp. 208–222, 2012.

established in vocational and advanced training over the last years and have a big potential for informal further vocational training [2].

1.2 Casual Game

There is a wide range of video games and there are lots of different aspects to classify them to genres. Focusing on interaction, Apperley [3] groups games into the genres simulation, strategy, action and role-playing. Crawford distinguishes skill-and-action games emphasizing perceptual and motor skills (e.g. combat, maze, sports, paddle, race) and strategy games emphasizing cognitive effort (adventures, D&D (role-playing), war games, educational and children's games, interpersonal games) [4]. Another aspect of games is the target group, ranging from casual to hardcore gamers [5]. To create a serious gaming framework for divergent target groups, we decided to focus on casual games. Casual games are known to be small games with a high potential for frequent gaming by people of various social and educational background. These games are characterized by simple and easy to learn rules and either by slowly increasing difficulty or a time limit combined with a high score list. They are not very time consuming and can be played occasionally, so they not only appeal to (hardcore) gamers but to the mass audiences [5]. Casual game can be assigned to a broad span of the above mentioned genres including puzzles, hidden object, time management, adventure, match 3, strategy, action and word games[1].

1.3 Collaborative Game

The gaming framework is intended for multiplayer games, but also supports single player games. Based on a constructivist view, the collaboration between the learners leads to knowledge sharing [6]. Furthermore, the opportunity to work together is supposed to be an incentive [7]. Since competition is an important incentive of gaming, the groups of two can compete by means of high score points against other players.

1.4 Game with a Purpose

To be flexible concerning the learning domain an ontology-based approach is used. The ontology may be easily exchanged to adapt the game to other domains. Furthermore, the ontology principally offers the opportunity to encode specific feedback for common misconceptions like it is often done in intelligent tutoring systems. Ontologies are usually incomplete, since it is nearly impossible to represent even a limited domain in full detail [8]. The traditional approach to ontology development is similar to software development containing phases of specification, conceptualization, formalization, implementation and maintenance [9]. Since this kind of ontology development is quite expensive, incentives for contributors are needed [10]. One approach is to use games for building, maintaining and aligning ontologies, addressing the

---

[1] http://www.bigfishgames.com/blog/the-next-casual-game-genre/
   (Big Fish Games is a big platform for casual games).

intrinsic motivation of playing games and concealing the actual work [10]. This approach is based on the idea of games with a purpose (gwap) [11], which are often used for human computing, i.e. by exploiting the human intelligence to facilitate tasks that are otherwise difficult or impossible to reach by computational means. Gwaps are typically collaborative casual games, in which pairs of players e. g. have to guess what words the partner is using to describe an image, decide if they are listening to the same tune or describe and respectively guess a secret word[2].

1.5   Overall Approach

The content of the games is provided as an ontology, so it can cover a wide range of domains. The ontology has very simple structure and thus can not only be created by knowledge engineers, but also directly by teachers. The game itself is kept simple with easy to learn rules and controls, and a session only takes a few minutes, so it can be used in lessons as well as in informal learning contexts. In the multiplayer mode two players play together, share their knowledge and are rewarded for agreements. In the singleplayer mode one player plays together with a bot which has the (true) knowledge of the ontology, so this mode can be used for self-testing. According to the assumption of incomplete knowledge, relations created by the players that do not occur in the knowledge base are not necessarily wrong, but possibly just missing, especially if a significant amount of players creates them. Therefore frequently occurring relations are very interesting, because either they can be used by knowledge engineers to enhance the ontology or they are typically misconceptions to be resolved by the teachers. Furthermore, the content of learning games is often provided by them, so the roles of the knowledge engineer and the teacher overlap.

Thus, apart from teachers being interested in getting information about common misconceptions to correct these, it is possible to use the so called "wisdom of the crowd" of the game players to enrich a pre-built ontology. The overall process, the involved parties and resulting artifacts are displayed in Fig. 1.

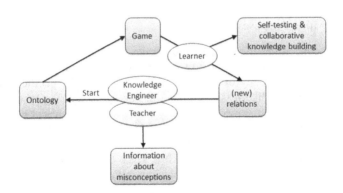

Fig. 1. Benefits of the game for learners, teachers and knowledge engineers

---

[2] http://www.gwap.com

2   Case Study

As a case study we created an ontology for the German food industry focusing on the domain of food safety and hazardous material regulations. The food industry in Germany is characterized by a high amount of workers without or with only a low level of formal qualification. While these workers are easily found and taught to perform the simple and often physically exhaustive tasks, there is a lack of employees with a higher qualification (e.g. skilled workers), who are able to use and control the complex machines and processes of the food production industry. Thus, the human resource managers try to train some of the lowly qualified to a higher qualification level to close the gap. This is not an easy task since these people often have a migration background and therefore language problems and/or they are not very motivated to learn because of various reasons, e.g. education is not an asset to them or their work is so exhausting that they are not ready to learn. The project Foodweb2.0 (funded by the German Ministry of Research and Education) aims at training the employees of the German food industry using two basic strategies: motivating employees for vocational training and performing education in collaborative, blended learning using Web2.0 technologies. Our framework is well suited to support this scenario as it combines both strategies. While playing the game the players have to remember facts and rules in the context of food safety and hazardous material regulations. Thus, it can be used in a corresponding course for training and recapitulation, not only in the class room, but also online at home.

3   Gameplay and Incentives

Matchballs is designed as a simple allocation game, in which the player creates statements by linking ("matching") concepts displayed as balls (see Fig. 2). A statement consists of two concepts linked by one of four predetermined relation types. The game can be played either as two player game or as single player game with a bot. Each pair of players sees the same game field which contains 15 concepts. The goal is to agree with the teammate on as many relations as possible in a given time. To agree on a relation both players have to create it. If they agree on a relation, they score points and get time bonuses. Players may see the relations of their teammates, but not the relation types. The initial relations of the player are visualized by hour glasses and the relations of the teammate by question marks. If the teammates agree on a relation, this relation is displayed by stars, if they disagree, it is displayed by red x.

In our case study we use the knowledge domain of food safety and hazardous material regulations, which is an important topic of further education in the German food industry. The considered concepts are specific situations, actions, dangerous substances and edibles, which can be linked by using the four semantic relations "is similar to", "is more general than", "results in" and "then you may not".

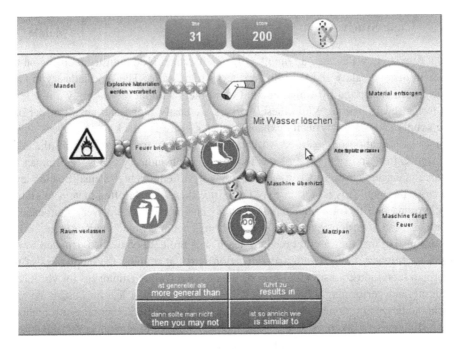

Fig. 2. Matchballs user interface

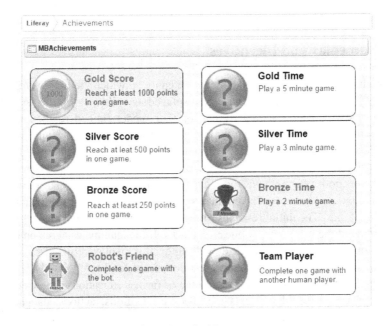

Fig. 3. Diplay of achievements

Feasible statements are for example:

- <Chicken><is similar to><turkey hen>
- <Machine overheats><results in><fire danger>
- <Oil starts burning><then you may not><extinguish the fire with water>

There are several incentives for playing the game considering different types of players: For competitive players there are high scores and time bonuses, which are a well-known and often used incentive (e.g. cf. [12]) since early arcade games. Furthermore, players can collect "achievements", which are trophies for solving certain predefined tasks. In Matchballs achievements are for example titles for playing a given number of games with another player ("team player") or with the bot ("robot's friend"), or cups for gaining certain amounts of points or extending the game for certain time spans (see Fig. 3). Achievements are a more recent kind of incentive often used in modern console games. They not only address competitive players, but also people with collector's passion, who want to unlock the full set of obtainable awards. While competitive players will tend to play against the bot to be not dependent on the team-mate, for team players the possibility to play together with another human is an incentive of its own.

4   Architecture

The Matchballs system is based on a multi-agent architecture. "The concept of multi-agent architectures relies on the idea that a collection of autonomous processes, called agents, can achieve intelligent problem-solving behavior by coordinating their knowledge, goals, skills, and plans" [12]. This kind of architecture is especially useful for systems which integrate different processing and reasoning methods and have the possibility to divide the problem solving knowledge into independent pieces [12]. Furthermore, multi-agent system can be easily extended. These characteristics fit our purposes very well. While there are currently only few analysis agents, there are lots of methods, e.g. from statistics, data mining or information retrieval, which could be used for further analysis integrating different theories. Furthermore, the results of our first tests (see section 5) could be used for the development of further agents.

The central game server is based on a tuplespace implementation called SQLSpaces [13]. Tuplespaces are inspired by blackboard architectures [15], which are characterized by a data-driven approach. Their general principle is to have no direct communication between the agents. These agents communicate by writing and reading messages on and from the blackboard (i.e. the Tuplespace). These messages consist of tuples made of primitive data types (integer, characters, booleans) and strings. This allows for a programming language heterogeneous approach, since these data types are available in almost all programming paradigms and languages. A single Tuplespace server may contain several tuplespaces used to divide the data stored in the server into logic or semantic units.

The Matchballs architecture distinguishes four different categories of tuplespaces: the Coordination Space, the Game Spaces, the Intermediate Space, and the Ontology Space (see Fig.4). The Coordination Space is used to conduct the matchmaking between two human players or to start a single player game. The GameClients register at the Coordination Space to announce their availability for a new game session and retrieve the information about the Game Space they have to connect to. In our case study the GameClients reside in a Liferay[3] portal, which is the JSR 2.0 portlet-oriented enterprise portal that is used for the Foodweb2.0 web presence. The Game Client is implemented only using HTML5 and JavaScript to cover most of the current browsers on the major operating systems including those that have no flash support. Furthermore, this allows for every HTML5-supported content inside of a Matchball. Although currently only images and text are used in a ball, video and audio files could also be played, e.g. for process sequencing tasks.

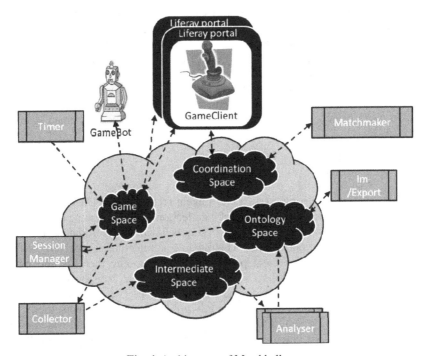

Fig. 4. Architecture of Matchballs

Currently all agents are implemented in Java, but some of the coming analysis agents conducting advanced analyses on the interaction data will probably be written in prolog, since this will be adequate to the analysis problem and the SQLSpaces implementation has also a prolog client.

4.1 The Game Space

Each game session has its own Game Space, to which either two human players put together by the Matchmaker agent through the Coordination Space for multiplayer games or a human player and a GameBot for single player games are connected. The

---

[3] http://www.liferay.com/en/

Game Space holds all necessary information for a Matchballs game. That means the Game Space consists of an excerpt of the ontology space, the current timer as set by the Timer agent, the current score, and the links made by the players as well as its assessment by the Session Manager. The GameBot has access to the whole information stored in the ontology excerpt, i.e. it is aware of the complete knowledge that is represented in that ontology excerpt. Thus, all associations made by the GameBot are correct assuming that the ontology is adequately modeled. The excerpt from the ontology is created by the Session Manager agent. It takes care that there is always a minimum of possible relations between balls in the beginning. It also detects concordances of the two players with respect to the links between balls made by each player. If an agreement on an association is detected by the Session Manager, i.e. both players' GameClients wrote exactly the same tuple representing a link between two balls to the Game Space, the Session Manager updates the score tuple of the current Game Space and writes an assessment tuple for this link into the Game Space. A game ends if the time is up. Afterwards the links made by the players (regardless if it was a multiplayer or single player game) are collected by the Collector and put into the Intermediate space for further inspection. The Game Spaces are discarded in a productive environment to save space. At the moment the Collector just counts the occurrence of the specific relations made by the players and stores or updates the amount in the Intermediate Space.

4.2   The Intermediate Space

The Intermediate Space is used for analyses. At the moment only very basic analyses are conducted. There is a threshold of at least five different players linking two concepts with the same association type (not present in the ontology yet) to propose this association for inclusion into the ontology. Another basic analysis looks for common misconceptions by the players, i.e. the players frequently contradict an association in the ontology by using another association type than the one represented in the ontology. Both analysis results are specifically marked in the Ontology Space.

4.3   The Ontology Space

The Ontology Space holds a tuple representation of the ontology. Every concept and relation is represented by a tuple. The markers set by the analyses agents are also represented in the ontology. The ontology design is based on SKOS (Simple Knowledge Organization System) [16]. As the acronym states SKOS is quite a simple representation mechanism, but it is sufficient for our game. In our case study, we distinguish only four association types, easily derived from SKOS-relations and we have few concepts/classes and many individuals/instances, which can be categorized in SKOS's concept schemes. The restrictions on the ontology are caused by the spirit of the game. Since it shall be a casual game for low qualified people a huge, sophisticated system of associations and concepts/classes would be misleading. With respect to the use of the game as an ontology enrichment game, it is also not feasible to have a complex representation scheme. Furthermore, SKOS is suitable for teachers to express their domain knowledge without much help from ontology engineers. Last but not least there is a plugin for Protégé for SKOS. Thus, we can use a popular editor for ontology creation, inspection and refinement. The Import-Export agent takes care of the proper translations of the tuple format to SKOS (OWL format) and vice versa.

The ontology used in our evaluation (see below) consists of 191 individuals connected with 91 associations distributed on the different association types. The structure of the ontology is displayed in Fig. 5.

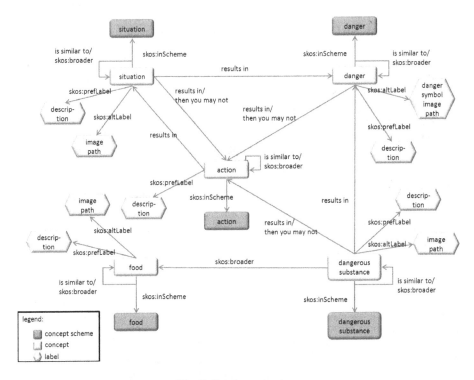

Fig. 5. Ontology structure

5   Evaluation

The Matchballs game has been evaluated with a class of 18 students at the Academy of Sweets in Solingen, Germany. Since there were only six laptops available for conducting the experiment, after a short plenary introduction of the game, the class was divided into three groups of six students. The students of each group had a timeslot of five minutes for playing as many games as possible and were encouraged to start with a single-player game for learning the controls and then perform at least one game with a human partner. After playing the game they had to complete a questionnaire and at the end there was a short plenary discussion.

The subjects consisted of five females and thirteen males, sixteen were German native speakers. They rated themselves to have high knowledge in the fields of safety at work and danger symbols (median of five on a scale from one to six) and also some knowledge of hazardous materials (median of four on a scale from one to six). 66.67 % declared to have participated in a course on safety at work and 55.56 % on a course on hazardous materials. Thus, the participants are considered to have at least basic knowledge of food safety and hazardous materials and to represent a group re-capitulating their knowledge on these topics.

Together the subjects created 155 different relations, most were only created by one user, but there were also relations created by up to seven different participants (see Fig. 6). The most often created relations are displayed in Table 1.

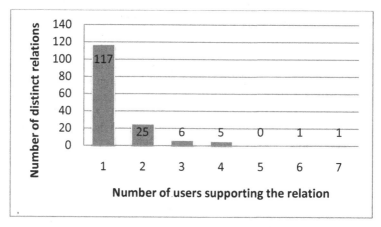

Fig. 6. Histogram of relations

Table 1. Relations support by the biggest amount of different users (translated to English)

concept	relation type	concept	support	assessment
pork	is similar to	beef	7	correct
pear	is similar to	apple	6	correct
switch on machine	is similar to	switch on comput-er	4	correct
barley	is similar to	rye	4	correct
machine uses laser beams	results in	danger by optical radiation	4	correct
machine shows functional disorder	results in	switch of machine	4	correct
radioactive material escapes	results in	radioactivity	4	correct
close the window	results in	draft at place of work	3	incorrect
danger of explosion	results in	explosive mate-rials are processed	3	incorrect
spelt	is similar to	barley	3	correct
chicken	is similar to	turkey hen	3	correct
machine is on high-voltage	then you may not	splash with water	3	correct
machine overheats	results in	fire danger	3	correct

The higher the amount of users supporting a relation, the higher is the probability that the relation is correct (precision) (see Table 2). A frequent error lies in choosing the wrong direction of the relation, around 26 % of the overall relations are correct except for the direction; if not all relations but the ones with a minimal support higher than one are considered, the percentage is even higher (see Table 2). These errors can be considered as careless mistakes instead of real misconceptions. Because of these mistakes, it is not amazing that the undirected relation "is similar to" has a better overall precision (0.53) than the other directed relation types. The relation "is more general than" is chosen most seldom (12.9 %) and has the worst overall precision (0.3). In the questionnaire 83.3 % of the subjects stated that they did not understand the meaning of this relation. The discussion in the plenum showed that flipping the relation and labeling it "is more specific than" would lead to a much better understanding, thus the relation was exchanged in the current version of the game client. With respect to the other relation types the subjects stated that they could comprehend them (94.4 % the relations "is similar to" and "then you should not" and 88.9 % "results in").

Table 2. Comparison of minimal support for considered relations, precision of the relations and percentage of errors based on wrong direction of the relation

minimal support	precision	errors based on wrong direction
4	1.00	none
3	0.85	100,00 %
2	0.68	58,33 %
1	0.40	25,80 %

The frequent correct relations having a support of at least three (see Table 1) already are in the ontology. This is due to the initial composition of the playing field containing two-thirds of concepts being linked in the ontology and only one-third of concepts being not linked with the other selected concepts (see section "Architecture"). But these relations only cover 17.7 % of the correct relations created by the users. Thus, the relations with high support have a high precision, but a low recall. 75.8 % of the correct relations already are in the ontology, 4.8 % of the correct new ones are supported by two and 19.4 % by only one user. The low support of the correct new concepts on the one hand can also be explained by the initial composition of the playing field and on the other hand by the limited number of games played in the experiment.

The "wisdom of the crowds" approach is often criticized arguing that expert contributions would be enough. In our case, the four "best" students (22.2 %) who created the biggest amount of correct new relations could only provide 46.67 % of the overall number of correct relations.

The users show different profiles of creating relations considering the number of different relations created, their precision, the number of correct relations missing in the ontology and the number of errors based on direction issues (see Table 3). Each user averagely created twelve relations, of which six (0.51 %) were correct, one

(8.3 %) was new and correct and ca. two (15 %) were wrong because of direction problems. There is no correlation between these variables. Partitioning the students into two clusters using the well-known kMeans algorithm [17] (see Table 4), reveals one small group of students (22.2 %), who create a number of distinct relations far below average, are a bit more precise and make no direction mistakes and a larger group, who create a bit more distinct relations than average, are a bit less precise than the other group and making much more direction errors.

On the whole the students had fun playing the game and experienced it as motivating (each variable had a median of four on a scale from one to six). 77.8 % stated they would like to play the game again and averagely stated they would play it once to several times a month. The complexity is perceived as medium (median of 3.5 on a scale from one to six), which is an indicator that the game is neither overstraining nor boring and hence appropriate.

Table 3. User profiles

User	Distinct relations/ activity	Precision/ quality	New correct relations/ innovativeness	Direction mistakes/ sloppiness
A1	12	0.50	8 %	17 %
A2	4	0.25	0 %	25 %
A3	17	0.47	12 %	12 %
A4	23	0.43	4 %	22 %
A5	14	0.57	7 %	29 %
A6	8	1.00	13 %	0 %
B1	6	0.33	17 %	0 %
B2	7	0.29	0 %	0 %
B3	10	0.60	0 %	30 %
B4	12	0.50	0 %	17 %
B5	8	0.50	13 %	25 %
B6	11	0.55	0 %	18 %
C1	8	0.63	13 %	13 %
C2	16	0.56	13 %	31 %
C3	18	0.67	6 %	6 %
C4	0	-	0 %	0 %
C5	26	0.27	8 %	8 %
C6	18	0.61	22 %	6 %

Table 4. Cluster centers

	Cluster 1 (14 students)	Cluster 2 (4 students)
Distinct Relations	14.07	5.25
Precision	0.49	0.51
Innovativeness	0.07	0.07
Sloppiness	0.14	0.00

6   Discussion

The evaluation results show that the Matchballs game was perceived as a casual game that is addictive enough to encourage the learners to try another game to improve their score. Based on the data generated during the game sessions we were able to identify 17 relevant associations that were not represented in the initial ontology. These associations were integrated into the ontology by our knowledge engineers. Thus, we could also indicate that our approach of using a learning game also as "game with a purpose" is feasible. In this way the game may be seen as self-extending with respect to closing gaps in the ontology. At the moment the game can only be used for adding new associations in the ontology.

In the future we plan to use the game to acquire new knowledge fields in a class session. The teacher may introduce a set of new concepts into an ontology which are not connected at all. The students are asked to play the game so that they will connect the new concepts with each other as well as with the old ones, therefore integrating them into the existing ontology. In a way the game may be viewed as a concept map creation game when used as a multiplayer game, where the players create a shared concept map. Concept maps are successfully used as learning tool for linking existing and new knowledge as well as for evaluation and identifying valid and invalid ideas of students [18]. Although, the effect of collaborative relation creation for learning is ambiguous [19, 20, 21, 22] and seems to depend on additional factors, the collaboration clearly does not impair the learning effect. Thus, a positive effect for the individual learner can be expected and the collaborative mode additionally may cause better performance and an enrichment of the underlying ontology.

If the game is played in single player mode, the game may still be used as an advanced vocabulary trainer. In spite of playing the game individually the students still collaborate indirectly. Accordingly, teachers can apply the game to get an overview of typical misconceptions of the group but also of single students.

The introduction of a high score table may raise the question, if it causes any friction among work or learning mates. As already mentioned above high score tables target people who are motivated by competition and may lead to unintended results (see [12]), nonetheless using well designed awards, like it is possible with Matchball, can be used to keep competition on a reasonable level. In the context of the Foodweb2.0 project there have already been several requests by teachers and students for transferring the game to further knowledge domains. We will try to incorporate these domains and enhance these ontologies with specific feedback on the relations made by students. The feedback will be given at the end of the game. For the multi-player scenario there will be a feedback about the existence of their relation in the ontology. In single-player scenarios the feedback hints at possible misconceptions automatically based on information directly represented in the ontology for particular error types and exploiting the semantic ontology structure for the generation of generic feedbacks like it is done in intelligent tutoring systems.

Acknowledgments. We would like to thank Christopher Charles, Peter Horster, Dominik Kloke, Carolin Pohl and Carsten Wieringer, who implemented the Matchballs game during a student project.

References

1. Michael, D., Chen, S.: Serious Games: Games that educate, train, and inform. Thomson Course Technology PTR, Boston (2006)
2. Marr, A.C.: Serious Games für die Informations- und Wissensvermittlung. Bibliotheken auf neuen Wegen, vol. 28. Dinges & Frick GmbH, Wiesbaden (2010)
3. Apperley, T.H.: Genre and game studies: Toward a critical approach to video game genres. Simulation & Gaming 37(1), 6–23 (2006)
4. Crawford, C.: The Art of Computer Game Design. Osborne/McGraw-Hill, U.S. (1982),
   `https://sakai.rutgers.edu/access/content/group/`
   `af43d59b-528f-42d0-b8e5-70af85c439dc/reading/crawford_1982.pdf`
5. Kuittinen, J., Kultima, A., Niemelä, J., Paavilainen, J.: Casual games discussion. In: Proceedings of the 2007 Conference on Future Play, pp. 105–112. ACM, New York (2007)
6. Cohen, E.G.: Restructuring the classroom: Conditions for productive small groups. Review of Educational Research 64(1), 1–3 (1994)
7. Johnson, R.T., Johnson, D.W.: Using cooperative learning in math. In: Davidson, N. (ed.) Cooperative Learning in Mathematics: A Handbook for Teachers. Addison-Wesley (1990)
8. Hepp, M.: Possible Ontologies: How Reality Constraints the Development of Relevant Ontologies. IEEE Internet Computing 11, 90–96 (2007)
9. Pinto, H.S., Martins, J.P.: Ontologies: How can They be Build? Knowledge and Information Systems 6, 441–464 (2004)
10. Siorpaes, K., Hepp, M.: Games with a Purpose for the Semantic Web. Journal IEEE Intelligent Systems 23(3), 50–60 (2008)
11. von Ahn, L.: Games with a Purpose. Journal Computer 39(6), 96–98 (2006)
12. Toups, Z.O., Kerne, A., Hamilton, W.: Motivating Play through Score. In: ACM Computer Human Interaction 2009 Workshop on Engagement by Design (2009)
13. Kipper, B.: Interlocking Multi-Agent and Blackboard Architectures. In: Pinto-Ferreira, C., Mamede, N.J. (eds.) EPIA 1995. LNCS, vol. 990, pp. 371–375. Springer, Heidelberg (1995)
14. Weinbrenner, S., Giemza, A., Hoppe, H.U.: Engineering Heterogeneous Distributed Learning Environments Using Tuple Spaces as an Architectural Platform. In: Proceedings of the 7th IEEE International Conference on Advanced Learning Technologies (ICALT 2007), Los Alamitos, CA, pp. 434–436 (2007)
15. Erman, L.D., Lesser, V.R.: A Multi-Level Organization for Problem Solving Using Many Diverse, Cooperating Sources of Knowledge. In: Proceedings of the Fourth International Joint Conference on Atrificial Intelligence, Tbilisi, Russia, pp. 483–490 (1975)
16. Miles, A., Bechofer, S.: SKOS Simple Knowledge Organization System Reference (2009),
   `http://www.w3.org/TR/2009/REC-skos-reference-20090818/`
17. MacQueen: Some methods for classification and analysis of multivariate observations. In: Proceedings of the 5th Berkeley Symposium on Mathematical Statistics and Probability, pp. 281–297 (1967)
18. Novak, J.D., Canãs, A.J.: The theory underlying concept maps and how to construct them. Technical Report IHMC CmapTools 2006-01 Rev 01-2008, Florida Institute for Human and Machine Cognition (2006),
   `http://cmap.ihmc.us/Publications/ResearchPapers/`
   `TheoryUnderlyingConceptMaps.pdf`

19. Esiobu, G., Soyibo, K.: Effects of concept and vee mapping under three learning modes on students' cognitive achievement in ecology and genetics. Journal of Research in Science Teaching 32(9), 971–995
20. Chung, G.K.W.K., O'Neil Jr., H.F., Herl, H.E.: The use of computer-based collaborative knowledge mapping to measure team processes and team outcomes. Computers in Human Behavior 15(3-4), 463–493 (1999)
21. Chinn, C.A., O'Donnell, A.M., Jinks, T.S.: The structure of discourse in collaborative learning. Journal of Experimental Education 69(1), 77–97 (2000)
22. Chiu, C.-H., Huang, C.-C., Chang, W.-T.: The evaluation and influence of interaction in network supported collaborative Concept Mapping. Computers & Education 34(1), 17–25 (2000)

Towards a Monitoring-Aware Design Process for CSCL Scripts

María Jesús Rodríguez-Triana, Alejandra Martínez-Monés,
Juan Ignacio Asensio-Pérez, and Yannis Dimitriadis

GSIC-EMIC, University of Valladolid, Spain
{chus@gsic,amartine@infor,juaase@tel,yannis@tel}.uva.es
http://www.gsic.uva.es

Abstract. Scripting and monitoring are two well-known strategies to enhance collaboration in CSCL settings. Teachers are incorporating them increasingly into their practice, however it is not common to find both of them aligned. We are working on the definition of a learning design process that takes monitoring issues into account and leads to better and more efficient monitoring when the scripts are put into practice. Moreover, if the learning design is based on patterns, the information given by these patterns can help to shape this enhanced design process. This paper presents a pilot study where a participatory design approach was followed. The first author and a teacher co-designed a CSCL situation in higher education based on the Jigsaw pattern. The analysis of the co-design process gave us a first structure of the data to be considered in monitoring-aware learning designs and a set of measures for enhancing monitoring at design-time.

Keywords: CSCL, learning design, scripting, monitoring, collaborative learning flow patterns.

1 Introduction

Scripting and monitoring are two strategies to shape group interactions in Computer-Supported Collaborative Learning (CSCL) scenarios [1] [2]. Scripting focuses on structuring the learning scenario, providing students with a set of instructions that guide potentially fruitful collaboration. It is performed at design-time, before the learning situation starts. Monitoring plays a relevant role in the regulation providing awareness information during the enactment of the learning situation [3].

It has been argued that monitoring could be more efficient if its requirements are considered at design-time [4] [5], as well as it has been done with the evaluation or the assessment [6]. In this direction, we are working on the definition of a *monitoring-aware design process*, where the specific characteristics of the learning design (constraints [7] [8]) guide the configuration of the monitoring process and where the requirements posed by monitoring are considered at design-time. Our work aims at exploiting the mutual relationships between monitoring and design, in order to improve both.

It is known that modeling potentially effective CSCL scripts is a difficult task (especially for non-expert designers), and also has been proven that the use of patterns that reflect good practices in structuring collaborative learning is helpful [9]. A particular type of CSCL scripting patterns are the *Collaborative Learning Flow Patterns*

V. Herskovic et al. (Eds.): CRIWG 2012, LNCS 7493, pp. 223–236, 2012.

(CLFPs), which capture the essence of well-accepted techniques for structuring the flow of collaborative learning activities [10].

In a previous work [11], we explored whether the use of CLFPs could be useful to guide monitoring. We found initial evidence that being aware of the pattern used in a script increases the opportunities of detecting critical situations when the script is put into practice. This way, monitoring process is more efficient, since the critical points are identified in advance, and monitoring can focus on detecting them, instead of on modeling the whole collaborative process.

However, the information provided by the CLFP is not sufficient. Other data derived from the specific characteristics of the learning situation is also needed to inform monitoring. In order to move forward in the definition of the aforementioned monitoring-aware design process, we set up a new pilot study, reported in this paper, with the intention of identifying the elements that should be considered in this process. The complexity of the envisioned process, and the mutual dependencies between design and monitoring call for a co-design process [12] [13]. We adopted this approach, with the teacher taking the role of expert "learning designer" and the researcher the role of "monitoring" expert. The requirements of both were considered and combined, in order to produce a "monitorable script" [11]. The goal was twofold: to identify the elements that would become part of a monitoring model on which to base the process, and to gather evidence on the usefulness of the proposed process from the teacher's point of view.

The work reported in this paper is aimed at answering the research question about which are the aspects that should be considered at design-time in order to monitor the learning scenario. This paper describes how the co-design process helped to identify these aspects. The paper also includes the main reflections from the participant teacher, who evaluated positively the monitoring process that was followed during the study.

The remaining of this paper is structured as follows: Section 2 describes the overall approach and previous work done by the authors towards this end. Section 3 presents the pilot study, and finally, conclusions and future work are summarized in Sect. 4.

2   General Approach

As mentioned beforehand, previous research has pointed out that synergies may appear when monitoring and design are aligned. On the one hand the design would benefit from taking into account the especial requirements posed by monitoring, and on the other hand, the integration in the design of the monitoring issues could help to obtain results better tailored to the teacher's needs.

However, no integration of monitoring has been observed into mainstream CSCL practices. Martínez et al. [4] classify the integration problems into two types: those caused by technological reasons and those that depend on the decisions taken during the design of the learning scenario.

Regarding the technological challenges, the main issues are related to the data gathering, interpretation, and integration. These obstacles increase when the technological context is heterogeneous and decentralized [14]. In these settings, it is necessary to process and take into account the information of distinct data sources in order to obtain a general and realistic view of the learning activities. In [15] we delved into the problem of data gathering in technologically decentralized learning environments for monitoring purposes. A solution was proposed to add monitoring functionalities to an existing architecture devoted to integrate virtual and personal learning environments (VLEs and PLEs) with external tools, named GLUE! (Group Learning Uniform Environment) [16]. Initial evidence was obtained on the capabilities of the proposed architecture to gather relevant information about the users' actions during the learning process.

In order to guide the configuration of the monitoring process, it is necessary to identify the constraints [7] [8] of the learning design that must be accomplished during its implementation. To address this issue, we analyzed the constraints of pattern-based CSCL scripts, in particular those based on CLFPs. From this analysis, we proposed a method to get an automated and higher level view about the evolution of the learning process by combining monitoring and pattern-based scripting. These ideas were tried out in an authentic learning scenario, and we found evidences that support that being aware of the pattern used in a script increases the opportunities of automatically detecting critical situations while the script is put into practice [11]. However, in this proposal the teacher was not involved in the configuration of the monitoring.

In summary, in our previous work we have tackled with the problems of the integration of tools and data at a technological level, and of how the CLFP constraints can help to shape a more efficient monitoring process. However, in order to achieve our goal of defining an enriched learning design process considering the requirements posed by monitoring, we still needed to gain insight on the factors that influence this process. We needed to identify the elements of a monitoring model that would become the base for the systematization of the monitoring-aware learning design process. This is the main goal of the study reported in this paper.

3 Identifying the Elements That Guide Monitoring: An Exploratory Study

This section reports the study that was set up in order to gain insight into the elements on which to sustain the envisioned monitoring-aware learning design process previously discussed. In this process, teacher and researcher followed a co-design approach of the learning and monitoring processes, thus ensuring that the needs of both were taken into account. Later on, the resulting monitoring-aware script was put into practice using the aforementioned GLUE! architecture. The participants' actions were monitored in order to test whether the overall design was being accomplished as expected.

This section is structured as follows: first, we present briefly the main characteristics of the case; then, we explain how the monitoring issues were taken into account during the co-design process, as well as their impact on the enactment; and finally, the section ends with the discussion of the results obtained from the case study.

3.1 Context and Methodologies of the Study

The case study presented in this section was developed from February 17th to March 9th, 2012, and took place within a course on "Learning methods" of a Preservice Master's Degree in Secondary Education, at the University of Valladolid (Spain), with 14 students attending the course. During this course, students had to analyze different teaching and learning methods. In order to help them in the understanding and internalization of these topics, they were asked to elaborate a poster with their choices about the most suitable learning methods for a concrete learning context. To elaborate this poster, students worked in a blended CSCL setting, interleaving *face-to-face* with *distance activities mediated by ICT (Information and Communication Technology) tools.*

The collaboration script implemented a *Jigsaw* CLFP. According to the definition of jigsaw given by [17], this pattern is especially intended for contexts in which several small groups are facing the resolution of the same problem, and typically, the problem to be solved is complex and can be easily divided into sections or independent sub-problems. For such context, this pattern provides some guidelines (a collaborative learning flow and a schema for group structuring) devoted to promote the feeling that team members need each other to succeed (positive interdependence), to foster discussion in order to construct student's knowledge, and to ensure that students must contribute their fare share (individual accountability).

The jigsaw suggests a sequence of activities consisting of: an *individual phase* in which each participant works *individually* on a particular sub-problem; an *expert phase* where participants studying the same problem meet in an *expert group* for exchanging ideas, becoming experts in the section of the problem given to them; and finally a *jigsaw phase* where experts from each expert group meet in *jigsaw groups* to contribute with their expertise to solve the whole problem. The requirements that must be satisfied to accomplish this pattern are summarized in Table 1. These requirements are taken from the CLFPs analysis described in the previous Section, where CLFPs where studied in order to identify the critical points that could guide the monitoring process [11].

With this case study we aimed at illuminating three issues that would help us to gain insight into our research goal. These issues were: "in which ways the consciousness about the CLFP constraints modifies the learning design?", "which are the parameters that model the monitoring configuration?" and, in a CSCL scenario characterized by the integration of ICT tools and the combination of face-to-face and distance activities, "which are the required conditions for collecting relevant information?".

In order to answer these questions, we found a co-design process between teacher and researcher the most suitable way to obtain a monitorable CSCL script and, at the same time, identify and analyze the relevant decisions made during its development. Then, the script was put into practice, and the results obtained were compared to those expected. These results were triangulated with data coming from observations carried out by the teacher during the face-to-face sessions; two questionnaires handed to the students about their work in groups; and several interviews to the teacher during and after the learning situation. This triangulation allowed us to verify the validity of the monitoring results and helped us collect the feedback of the teacher.

3.2   The Co-design Process

As aforementioned, during this phase teacher and researcher worked together on a design that took into account both the pedagogical and monitoring needs. Throughout this process, we pursued to identify different dimensions and attributes of the learning design that may guide a design-based monitoring process.

The co-design process consisted of two cycles. First, the teacher designed the learning situation following the guidelines given by the pattern-design process. The researcher contributed with her knowledge on the pattern, observing how the decisions taken by the teacher influenced monitoring, and intervening where necessary to ensure a better process. In the second cycle, both teacher and researcher worked together, exploring additional ways on which the script could be improved in order to obtain the information needed to detect the critical situations given by the pattern. A summary of the main decisions made both parts of the co-design process have being summarized in Table 2. Italicized text is used for the decisions made in the second cycle of the study in order to improve the monitoring process.

First Cycle: The Pattern-Driven Co-design. Focusing on the learning objectives, the teacher designed the scenario following a pattern-based learning design process. This process was supported by the authoring tool *Web Instance Collage* [18], which defines four main steps: (1) the definition of *objectives and prerequisites* of the learning design, (2) the *creation of groups* and the corresponding allocation of students, (3) the particularization of the *activity learning flow* and (4) the *provision of resources* (contents and tools). Here, we will focus on the last three ones (group creation, definition of the activity learning flow and provision of resources) as current aspects of interest from the standpoint of monitoring. A summary of the main decisions taken during these co-design cycles is presented in Table 2.

Throughout this process, focusing on the monitoring issues, the researcher informed the teacher about **constraints that should be accomplished according to the pattern** and the impact that they could have on the learning situation if they were not met. Table 1 summarizes the analysis of the jigsaw constraints.

The first of these steps, **group creation**, is of paramount importance for the success of the learning situation. The information obtained from the group structuring is very useful for the monitoring labor, because it *gives information about the expected structures of interactions in a given activity*. In our case, the design was based on a jigsaw CLFP, group formation consisted on distributing students in concrete *jigsaw* and *expert groups*. As mentioned in Table 1, each expert group had to contain at least one member of each jigsaw group, and viceversa, each jigsaw group had to include at least one member of each expert group. Expert groups have to include as many members as the number of sub-problems identified. At design-time, the designer can be advised to replicate these "roles" of expert and/or member of a jisgsaw group, in the cases where the absence of one or several students is known in advance. This happened in the case being described. After a query to the students, the teacher decided to replicate one of the experts in a jigsaw group, because one of the students was not expected to assist to the course, in spite of being enrolled in it. From the 14 students, 3 jigsaw groups and 4 expert groups were defined. 12 students were assigned to these groups in order to ensure the pattern constraints, and the other two were allocated to existing groups. These aspects are not normally possible to foresee, but will only appear when the activity is put into practice. As we will describe later in this section, the monitoring process is then the responsible of testing these critical points defined by the CLFP. This would, for example, help to detect whether a particular jigsaw group is lacking the contribution of one expert to it.

Table 1. List of constraints of the jigsaw CLFP. **X** represents that the restriction must be satisfied in that specific phase of the pattern (individual, expert and jigsaw).

Structuring constraints	Individual (individual)	Expert (collaborative)	Jigsaw (collaborative)	Description
group sizes		X	X	There must be enough participants to collaborate.
expert group sizes		X	X	The group sizes must be large enough to provide participants to each jigsaw group.
jigsaw group sizes			X	The group sizes must be large enough to gather experts from all areas.
no. of subproblems	X	X	X	There must be at least 2 subproblems but no more than the half of the class to allow for collaboration.
no. of expert groups	X	X	X	There must be at least one group of experts for each subproblem but no more than the half of the class to allow for collaboration.
no. of jigsaw groups			X	The number of jigsaw groups must be in accordance to the number of experts of each area.
group dependences			X	There must be experts of all areas in each jigsaw group.

The next design step was the **description of each activity** within the learning flow described by the CLFP. This was done by the teacher, who defined the concrete tasks that the students had to accomplish during the three phases of the jigsaw pattern. At the individual phase, each participant had to studied two learning methods out of an overall number of six. At the end of the phase they had to write an individual summary about the studied methods. During the expert phase, those students that had been working on the same methods joined into a expert group and designed collaboratively a concept map with the main ideas of each of the two methods they had studied. Once the expert phase finished, students were distributed in jigsaw groups. During the jigsaw phase, the students worked in their jigsaw groups, and the planned activities consisted on the elaboration of a poster where they had to choose two methods out of the six they had studied in the group, and justify their choice, discussing their suitability for the learning contexts they were working on. The poster and its presentation was evaluated by the rest of the classmates in an oral presentation at the end of the activity. The definition of the activities was complemented with decisions about their duration (with explicit starting and ending points), their interactivity type (face-to-face, through computers or blended, and their physical locations (inside and/or outside the classroom). From the monitoring point of view, being aware of which activities are to be carried out individually or in groups, (and in which groups), gives information about which evidences should be gathered in each phase. Besides this, the time limits are needed to narrow the period of the analysis; and the combination between the interactivity type and the location gives information about which evidences are applicable and potentially useful (i.e., presence in a face-to-face activity in groups, or submission of a deliverable in an individual task, etc.) or not (i.e., it is not appropriate to monitor the number of individual accesses to a tool if only a (unique) group submission is expected at the end of the task). Besides this, some situations, such as face-to-face interaction in the classroom, need additional sources of data not provided by the tools, in order to register the presence or absence of students in that task.

Last step in the design process is the **provision of necessary resources**. Apart from the bibliography, the design required ICT tools for collaborative drawing and writing as well as on-line questionnaires. Then, the next step involved the searching of tools that satisfied the teacher's needs and at the same time allowed us to harvest data about the users' actions. On the one hand, the teacher posed the restriction of using *MediaWiki*[1] to centralize the access to all the resources and activities and, on the other hand, the researcher was interested on tools that provide data about the users' actions. Both of them agreed on using the *GLUE!* architecture [16] because it allows the integration of external tools on MediaWiki and besides, it facilitates the collection of information from the different technologies used in the learning scenario [15]. Initially, the teacher proposed to use *Text 2 Mind Map*[2], a web application for development of conceptual maps, and *Google Forms*[3] for the on-line questionnaires. However, since Text 2 Mind Map did not offer any information about user actions, it was substituted by *Dabbleboard*[4] by proposal of the researcher. From such technological context it was possible to detect who accessed in a specific moment to Dabbleboard or Google Forms, as well as the editions and uploads done by the users of MediaWiki. Being aware of the tools required for each activity allowed us to **focus the monitoring data harvesting on the significant sources**.

Second Cycle: Enriching the Design to Enhance Monitoring. Up to this point, the co-design process had been driven by the CLFP-based design approach. The teacher had followed the phases described in it, including some aspects that could improve monitoring, based on the knowledge that the researcher had on this topic. This approach had provided the teacher with a partial view about potential critical situations detected during the enactment. However, there was a need of going one step further, looking for new ways on which the design itself could be modified in order to better inform the monitoring process.

At this point the focus was on how the design could be modified in order to augment the information given by ICT tools, and help teachers detect the potential critical situations. On the one hand, we worked on **enhancing monitoring data sources** that

---

[1] http://www.mediawiki.org
[2] http://www.text2mindmap.com/
[3] http://www.google.com/google-d-s/forms/
[4] http://www.dabbleboard.com/

Table 2. Description of the activities included in the script. Italicized text is used for the elements that were added in the second cycle of the study in order to improve the monitoring process. (ITC - Inside the classroom / OTC - Outside the classroom).

Phase	Activity	Social level	Interactivity type	Physical location	Resources and tools for learners	Resources and tools for teachers
Individual	Individual study	Individual	Through computers	OTC	- Documentation on learning methods	
Expert	Individual summaries	Individual	Through computers	OTC	- A wiki page	*- Monitoring report* - Register of submissions
	Expert consensus	Expert groups	Blended	ITC & OTC	- A shared board (Dabbleboard) - A wiki page	*- Register of attendance* *- Monitoring report* - Register of submissions
	Workgroup report	*Expert groups*	*Through computers*	*OTC*	*- A questionnaire (Google Forms)*	*- Monitoring report* - Register of submissions
Jigsaw	Selection of methods	Jigsaw groups	Through computers	OTC	- A questionnaire (Google Forms)	*- Monitoring report* - Register of submissions
	Poster development	Jigsaw groups	Blended	OTC	- A wiki page	*- Monitoring report* - Register of submissions
	Peer review	Individual	Through computers	OTC	- Wiki pages	*- Monitoring report* *- Register of participation* - Register of submissions
	Posters presentation	Jigsaw groups	Face-to-face	ITC		*- Register of attendance* *- Register of participation* - Register of submissions
	Workgroup report	*Jigsaw groups*	*Through computers*	*OTC*	*- A questionnaire (Google Forms)*	*- Monitoring report* *- Register of submissions*
	Peer evaluation	Individual	Blended	ITC & OTC	- A questionnaire (Google Forms)	*- Monitoring report* - Register of submissions

could provide relevant information about the learning situation. Table 3 summarizes the informants that were identified (the technological support, the teacher herself and the students) depending on the physical location and interactivity type of the specific activity. As it can be observed, the interactivity type and physical location have an influence on which data sources can be used to get information about one activity. For example, activities being performed face-to-face outside the classroom can only be informed by the students themselves, while those mediated by computers inside the classroom can be informed by the data collected by the tools (ICT), by the teachers in their observations of the class and by the students themselves. Between these two extremes, other combinations of data sources exist, as shown in Table 3. It is important to note than in blended settings, there are many interactions that happen out of the scope of the technology and even of the classroom. Therefore, if these activities are to be monitored, additional monitorable data sources that capture these data are needed. We have called them **data gathering activities**, as their main function is to enable the collection of new data that will help to identify potential critical situations.

Table 3. Data sources needed for the monitoring of a collaborative activity depending on the interactivity type (face-to-face, computer mediated or blended) and the physical location (ITC - Inside the classroom and/or OTC - Outside the classroom).

	Face-to-face	Blended	Computer mediated
ITC	students & teachers	students & teachers & technologies	students & teachers & technologies
ITC & OTC	students & teachers	students & teachers & technologies	students & teachers & technologies
OTC	students	students & technologies	students & technologies

According to this, the learning design obtained in the previous cycle of the process was enriched with additional **data gathering activities**, in order to collect data from teachers and students (see text in italics in Table 2). For every activity fully or partially located in the classroom, the teacher programmed to *control the attendance and participation* in order to take into account what happened during the these sessions. And for those collaborative activities that happened out of the classroom, students were asked to fill out a *a form about the distribution of tasks in their groups* (named "workgroup reports" in Table 2).

In order to complement all these activities, and to make explicit the monitoring process, the design was enriched with additional **monitoring support activities** to be performed by the teacher. These activities are a means to remind teachers what they have to do to support the students during each phase of the activity (e.g. verify that all jigsaw groups submit their posters, check that every group receives feedback from at least one reviewer, control that in every group there are enough people participating, etc.). These supporting activities were informed by the workgroup reports and the *monitoring reports*, which collected the data gathered from the different tools integrated in the resulting learning platform, and from the attendance control forms filled out by the teacher during the activities that required this data.

Summary: Elements of Monitoring Identified in the Study. As a synthesis to this co-design process, we will summarize here the connections that emerged between learning design and monitoring. On the one hand, the monitoring needs of **data gathering** caused changes in the original learning design such as choosing *tools* that provided data about the users' interactions, identifying complementary *monitoring data sources* to avoid blind spots, and including additional *data gathering activities* for teacher and students to collect evidence about the learning process, and *monitoring support activities* to be performed by the teacher.

As a result of this process, we have identified three dimensions that influence the design of the monitoring process (see Table 4). In a pattern-based process, the first one is the **design pattern**, by means of the constraints that must be verified during the enactment to accomplish the pedagogical objectives. These constraints, namely the *collaborative learning flow* and the *group formation policies*, constitute one of the aspects that have to be informed by the monitoring. A second dimension is related to the specific **features of each activity**: the *time period* in which an activity is of interest to study, the concrete *resources* that will be analyzed, how students are expected to develop the activity at a *social level* (individually or in groups), the *interactivity type* (face-to-face through computers or blended) and the *location* (inside and/or outside classroom). The

Table 4. Compilation of parameters set by the pattern, the definition of each activity and the teacher's interests in the reported study

Pattern	Activity	Teacher's interests
Activity flow	Deadlines	Monitoring periods
Collaboration	Resources (tools, contents)	Relevant resources
Group formation policies	Social level	Actions to be monitored
	Interactivity type	Additional constraints
	Location	

third dimension configuring the monitoring process are **teachers**, who can tune the configuration of the monitoring according to their needs, specifying aspects such as when the monitoring report is to be received, the *resources that have to be analyzed* in each activity, which *student actions* are meaningful to better understand the learning process, and some *additional constraints* to be checked to ensure the viability of the learning situation, such as the dependences among activities not reflected in the pattern (i.e. availability of artifacts for a peer review activity).

3.3   Enactment of the Learning Situation

The script was put into practice in the context previously described. A monitoring report was sent to the teacher 15 minutes after the deadline of each activity. In most cases this report helped her to confirm that the students were following properly the script. Nevertheless, some unexpected events appeared during the enactment that helped the teacher to take regulatory measures. We describe them here, in order to illustrate the impact that monitoring had in improving the overall learning situation.

For instance, in three of the activities: *individual summaries*, *peer review* and *peer evaluation* (see table 2) there was no evidence of some of the students having performed their tasks. In these situations the teacher started by verifying the work done by the students, and in the cases where the problem was confirmed, she sent a reminder, extending the deadlines. These problems could not have been detected by the teacher without the monitoring report. A similar problem arose with the *workgroup report* carried out by expert groups, where no evidence of participation was registered by two of the groups. However, in this case the reason was due to a technological problem with the on-line questionnaires supporting the activity, that could be easily fixed on the fly, and the students could submit their answers on time.

Another issue happened during the *expert consensus*. In this case, the monitoring report informed the teacher that two of the groups had not submitted their deliverable on time. If confirmed, this problem would have become a critical situation in the enactment of the pattern, as the lack of these deliverables affected the upcoming jigsaw phase. However, reviewing the work done by the students, the teacher realized that the contributions had been submitted at erroneous pages of MediaWiki.

In this latter case, as well as in the previous one, although the problem detected was not due to a fault in the students' performance, monitoring helped the teacher to detect and solve them. Overall, in this case monitoring helped the teacher to confirm with almost no effort on her part that the students were performing as expected. In the cases where this did not happen, to detect the problem and solve it before it became a real breakdown in the activity.

3.4 Discussion

As mentioned in the Introduction, the main goals of this study were two: to identify the elements that would become part of a monitoring model on which to base the process, and to gather evidence on the usefulness of the proposed process from the teacher's point of view. For achieving these goals we proposed three research issues, which were presented in Sect. 3.1. We discuss the results according to them.

The first issue considers the impact that the awareness of the CLFP constraints had over the learning design. The teacher participating in this study was familiar with CSCL pattern-based activities, and used CLFP-based authoring tools (Web Instance Collage in this case). The co-design process showed that, in spite of these facts, she had not had in mind any of the parameters to be controlled in the script, neither was she conscious about the impact that violations of the pattern constraints might have had over the whole design. Once the teacher realized about the potential critical points of her design and the lack of data to inform them, she decided to include new activities for students and teacher in order to facilitate the data gathering from additional sources. Besides this, her awareness on the information provided by the ICT tools increased, and she became more conscious about the need of monitoring the learning situation during their enactment.

The second issue is referred to the identification of parameters that model monitoring. This pilot study has helped us to identify three dimensions that influence the configuration of the monitoring process: the design pattern, the features of each activity and the teachers. Each one of these dimensions poses a set of parameters that condition what, how and when each activity should be monitored (see Table 4). This information will help us to systematize the process and move forward on the definition of a monitoring-aware learning design process.

The third and last issue is related to the conditions required for collecting relevant information in a CSCL scenario characterized by the integration of ICT tools and the combination of face-to-face and distance activities. As reflected in Table 3, depending on the interactivity type and the physical location, data sources available may vary (students, teacher and technological support). Thus, the monitoring process can not be exclusively based on the information obtained from the technological support. In those activities developed inside of the classroom, the input from teachers and/or observers is crucial to complete the information given by the ICT tools (e.g. using attendance and participations registers), as well as the feedback given by students about their work out of it. Obviously, in order to facilitate these tasks, it will be necessary to provide teachers and students with tools that support the data collection (e.g. by means of on-line forms to report the students attendance or on-line questionnaires to collect the participants comments). The study reported in this paper has shown the feasibility of this approach.

Finally, we collected some impressions and feedback from the teacher's interviews carried out during and after the experience. Reflecting on the design process, the teacher perceived the integration of monitoring issues as another aspect of the design to be taken into account that helped her to enrich the script. Besides, the teacher got a perception of better control of the learning situation since she knew where to focus the attention during the enactment. In relation to the monitoring results, they helped the teacher to detect potentially critical situations, facilitating the regulation of eventualities. Moreover, she declared that the monitoring reports were much more accurate to the teacher needs (because of her involvement in the monitoring configuration) and closer to the real facts (due to the integration of several data sources). Finally, evaluating the whole process, the teacher confirmed that improvements and results outweighed the effort devoted for the configuration of the monitoring process and the data gathering.

4   Conclusions and Future Work

This paper presents a case study, carried out in a higher education learning scenario with the aim of identifying the elements that should be considered in a monitoring-aware learning design process. On account of the complexity of this process, and the mutual dependencies between design and monitoring, teacher and researcher worked together in a co-design process.

As a result of this process, we have identified three dimensions that influence the configuration of the monitoring process in pattern-based approaches: the design pattern, the specific features of each activity, and the teacher's choices about specific issues. From each one of them, a set of parameters that guide the monitoring process has been identified. This information will help us to systematize the monitoring-aware design process in the near future. Nevertheless, since just one researcher and teacher were involved in the co-design process, this study will be extended with other participants to avoid bias.

The study has also provided additional information about the design process itself. By means of reflecting on the pattern constraints during the design, the teacher has included additional data gathering activities in order to inform the critical points of the script, improving both the learning design and the monitoring. These positive results support our idea of working towards the definition of a process that include monitoring as another design dimension. The process could eventually provide design-time scaffolding for helping teachers to consider monitoring implications of their pedagogical decisions.

The monitoring results provided during the enactment have demonstrated once more to be helpful for facilitating the regulation tasks, as stated by [3]. The works reported in [19] [20] and [21], analyzing the scripts flexibility and adaptability, could benefit from our proposal. Once teachers know which constraints are not being satisfied in real time, it seems to be easier for them to address the issues hampering the learning situation. This benefit is even more remarkable in heterogeneous scenarios where several ICT tools are involved [14].

Future work lines include two main threads. On the one hand, the elements that influence monitoring identified in this study must be consolidated into a model; this model will be the base for the integration of monitoring issues into existing (or new) authoring tools, in order to generate monitorable scripts. This integration will entail providing teachers with information about the "monitoring" properties of the learning tools (what information is offered about the users interactions and how can it be retrieved), in order to facilitate an appropriate selection that satisfies both the pedagogical and monitoring needs. On the other hand, we have detected a clear need of providing teachers with tools that facilitate the data gathering about the learning context (for instance attendance or participation registers), offer visualization of the monitoring results, and allow the modification of the monitoring results with the teacher's knowledge about what is happening during the enactment (e.g. confirming submissions, modifying deadlines, etc.). This is a work line very close to the approaches followed in workflow systems [22].

Acknowledgements. This research has been partially funded by the Spanish Ministry of Science and Innovation (projects TIN2008-03-23, TIN2011-28308-C03-02 and IPT-430000-2010-054) and the Autonomous Government of Castilla y León, Spain (projects VA293A11-2 and VA301B11-2). The authors would also like to thank the rest of GSIC/EMIC Group at the University of Valladolid for their support and ideas.

References

1. Jermann, P., Soller, A., Lesgold, A.: Computer Software Support for CSCL. In: Strijbos, J.W., Kirschner, P., Martens, R. (eds.) What We Know about CSCL. and Implementing it in Higher Education. Computer-Supported Collaborative Learning, vol. 3, pp. 141–166. Kluwer Academic Publishers, Norwell (2004)
2. Dillenbourg, P., Hong, F.: The mechanics of CSCL macro scripts. International Journal of Computer-Supported Collaborative Learning 3, 5–23 (2008)
3. Prieto-Santos, L.P., Holenko-Dlab, M., Gutiérrez, I., Abdulwahed, M., Balid, W.: Orchestrating technology enhanced learning: a literature review and a conceptual framework. International Journal of Technololgy Enhance Learning 3(6), 583–598 (2011)
4. Martínez-Monés, A., Dimitriadis, Y., Harrer, A.: An interaction aware design process for the integration of interaction analysis in mainstream CSCL practices. In: Analyzing Collaborative Interactions in CSCL: Methods, Approaches and Issues, pp. 269–291. Springer (2011)
5. Lockyer, L., Dawson, S.: Learning designs and learning analytics. In: Proceedings of the 1st International Conference on Learning Analytics and Knowledge, LAK 2011, pp. 153–156. ACM, New York (2011)
6. Villasclaras-Fernández, E., Hernández-Leo, D., Asensio-Pérez, J., Dimitriadis, Y.: Incorporating assessment in a pattern-based design process for CSCL scripts. Computer Human Behaviour 25, 1028–1039 (2009)
7. Dillenbourg, P.: Over-Scripting CSCL: the risks of blending collaborative learning with instructional design. In: Three Worlds of CSCL. Can We Support CSCL?, pp. 61–91. Open Universiteit Nederland, Heerlen (2002)
8. Kobbe, L.: Framework on multiple goal dimensions for computer-supported scripts, Kaleidoscope Noe. Deliverable D29.2.1 (2005)
9. Conole, G., McAndrew, P., Dimitriadis, Y.: The role of CSCL pedagogical patterns as mediating artefacts for repurposing Open Educational Resources, Hershey, USA (2011)
10. Hernández-Leo, D., Villasclaras-Fernández, E., Asensio-Pérez, J., Dimitriadis, Y., Retalis, S.: CSCL Scripting Patterns: Hierarchical Relationships and Applicability. In: Proceedings of the 6th IEEE International Conference on Advanced Learning Technologies, ICALT, Kerkrade, The Netherlands, pp. 388–392 (2006)
11. Rodríguez-Triana, M.J., Martínez-Monés, A., Asensio-Pérez, J.I., Jorrín-Abellán, I.M., Dimitriadis, Y.: Monitoring Pattern-Based CSCL Scripts: A Case Study. In: Kloos, C.D., Gillet, D., Crespo García, R.M., Wild, F., Wolpers, M. (eds.) EC-TEL 2011. LNCS, vol. 6964, pp. 313–326. Springer, Heidelberg (2011)

12. Steen, M., Manschot, M., De Koning, N.: Benefits of Co-design in Service Design Projects. International Journal of Design 5(2), 53+ (2011)
13. Visser, F.S., Stappers, P.J., van der Lugt, R., Sanders, E.B.N.: Contextmapping: experiences from practice. CoDesign: International Journal of CoCreation in Design and the Arts 1(2), 119–149 (2005)
14. Voyiatzaki, E., Polyzos, P., Avouris, N.: Teacher tools in a networked learning classroom: monitor, view and interpret interaction data. In: Networked Learning Conference, Halkidiki, Greece (2008)
15. Rodríguez-Triana, M., Martínez-Monés, A., Asensio-Pérez: Monitoring Collaboration in Flexible and Personal Learning Environments. Interaction, Design and Architecture(s) Journal, special issue on: Evaluating Educative Experiences of Flexible and Personal Learning Environments 11(12), 51–63 (2011)
16. Alario-Hoyos, C., Wilson, S.: Comparison of the main Alternatives to the Integration of External Tools in different Platforms. In: Proceedings of the International Conference of Education, Research and Innovation, ICERI 2010, Madrid, Spain, pp. 3466–3476 (2010)
17. Hernández-Leo, D., Villasclaras-Fernández, E., Asensio-Pérez, J., Dimitriadis, Y.: Diagrams of learning flow patterns solutions as visual representations of refinable IMS Learning Design templates. In: Handbook of Visual Languages for Instructional Design: Theories and Practices. IGI Global (2007)
18. Villasclaras-Fernández, E.D., Hernández-Leo, D., Asensio-Pérez, J.I., Dimitriadis, Y., Martínez-Monés, A.: Towards embedding assessment in cscl scripts through selection and assembly of learning and assessment patterns. In: Proceedings of the 9th International Conference on Computer Supported Collaborative Learning, CSCL 2009. International Society of the Learning Sciences, vol. 1, pp. 507–511 (2009)
19. Dillenbourg, P., Tchounikine, P.: Flexibility in macro-scripts for CSCL. Journal of Computer Assisted Learning 23(1), 1–13 (2007)
20. Demetriadis, S., Karakostas, A.: Adaptive Collaboration Scripting: A Conceptual Framework and a Design Case Study. In: Proceedings of the 2nd International Conference on Complex, Intelligent and Software Intensive Systems, CISIS 2008, pp. 487–492. IEEE Computer Society, Washington, DC (2008)
21. Karakostas, A., Demetriadis, S.: Adaptation patterns in systems for scripted collaboration. In: Proceedings of the 8th International Conference on Computer Supported Collaborative Learning, CSCL 2009, Rhodes, Greece. International Society of the Learning Sciences, vol. 1, pp. 477–481 (2009)
22. Fu, X., Hu, J., Teng, S., Chen, B., Lu, Y.: Research and Implementation on CSCW-Based Workflow Management System. In: Shen, W., Yong, J., Yang, Y., Barthès, J.-P.A., Luo, J. (eds.) CSCWD 2007. LNCS, vol. 5236, pp. 510–522. Springer, Heidelberg (2008)

Author Index